RECAPTURING DEMOCRACY

RECAPTURING DEMOCRACY
Neoliberalization and the Struggle for Alternative Urban Futures

Mark Purcell

ROUTLEDGE
NEW YORK AND LONDON

First published 2008
by Routledge
270 Madison Avenue, New York, NY 10016

Simultaneously published in the UK
by Routledge
2 Park Square, Milton Park, Abingdon, Oxon OX14 4RN

Routledge is an imprint of the Taylor & Francis Group, an informa business

© 2008 Taylor & Francis

Typeset in Minion by 1007525233
HWA Text and Data Management, Tunbridge Wells
Printed and bound in the United States of America on acid-free paper by
Edwards Brothers, Inc.

Library of Congress Cataloging in Publication Data
Purcell, Mark Hamilton
    Recapturing democracy : neoliberalization and the struggle for alternative
    urban futures / Mark Purcell – 1st ed.
        p.   cm.
    Includes bibliographical references
    1. Urban economics   2. Neoliberalism   3. Democracy   I. Title
    HT321.P87 2008
    330.9173′2–dc22   2007027725

ISBN10: 0–415–95434–7 (hbk)
ISBN10: 0–415–95435–5 (pbk)
ISBN10: 0–203–93294–3 (ebk)

ISBN13: 978–0–415–95434–1 (hbk)
ISBN13: 978–0–415–95435–8 (pbk)
ISBN13: 978–0–203–93294–0 (ebk)

FOR EDWARD BENTON PURCELL,
in whose footsteps I follow.
I search for the right word, but there is none.

CONTENTS

ILLUSTRATIONS

TABLE

Acknowledgements

Townes Van Zandt used to claim that his song "Mr. Gold and Mr. Mudd" just came to him, as if from above, and all he did was write it down as fast as he could. Maybe. For the rest of us, and especially for a book like this, writing is a long-term process of accreting knowledge and then trying to articulate what you have built up in a reasonably coherent way. The source of almost all that knowledge is the wisdom of others. And so I want to acknowledge the many others who have contributed enormously to this book.

Intellectually, the book has several key inspirations. The idea of democratic resistance that lies at the center of the book's argument is built significantly out of the remarkable work of Chantal Mouffe, Ernesto Laclau, Nancy Fraser, Michael Hardt, and Antonio Negri. My account of neoliberalization owes an enormous amount to the pioneering research of David Harvey, Bob Jessop, Neil Brenner, Nik Theodore, Sallie Marston, Jamie Peck, Adam Tickell, Erik Swyngedouw, and Wendy Larner. And my conception of the idea of the right to the city, of course, draws extensively on the extraordinary Henri Lefebvre.

In addition to those main theoretical figures, the book has also been shaped in countless ways by my engagements with colleagues and students. Of central importance here have been my discussions with students in my seminar on urban democracy. Special thanks go to Kevin Ramsey, Matt Wilson, Dawn Couch, Keith Goyden, Marcus Riccelli, Tom Barnard, Gary Simonson, Tuna Ali, Brennon Staley, David Moore, and Diego Velasco.

Also important have been my colleagues at the University of Washington and in Seattle beyond. I want to single out what I have learned from Branden Born, Chris Campbell, Dennis Ryan, Ralph Coolman, Jan Whittington, B.J.

Cummings, Nancy Beadie, Don Argus, Sheila Valencia, Steve Herbert, Sarah Elwood, Katharyne Mitchell, Scott Miles, and Michael Brown. And I must make special mention of Walter Parker: his contribution as an intellectual, mentor, colleague, and friend are immeasurable.

I want to thank a broad and diverse intellectual community beyond Seattle. My committee at UCLA, Gerry Hale, John Agnew, and Mark Ellis, helped shape me and my project in very different but always stimulating ways. Gerry Hale's intellectual and political integrity are a model that I am trying to live up to. Ed Soja and Mike Davis were absolutely central to my development as a scholar. I also owe a great debt to my friend-colleagues from UCLA, especially Monica Varsanyi, Garth Myers, Joe Nevins, and for so very many things, Chris Brown. Also of infinite help not only in Los Angeles but also well beyond, has been Ron Eltanal. His vibrant cynical wonderment is a thing to behold.

Beyond UCLA, I have been privileged to learn from and be greatly helped by many people, both professionally and intellectually, including Bob Lake, Kevin Cox, Andy Jonas, Pauline McGuirk, Mike Heffernan, Jamie Peck, John O'Loughlin, Bob Beauregard, David Wilson, Lynn Staeheli, Sallie Marston, Michael Hardt, John Forester, and especially Neil Brenner. I am also forever grateful to Eugene McCann and Deb Martin, both for their friendship and for helping me understand I was not alone in my fascination with urban politics.

I also want to acknowledge a host of people in both Seattle and Los Angeles who were the focus of the empirical research. I will not mention them by name so their anonymity can be preserved, but they should all know that I truly appreciate their patience, kindness and willingness to help. I *can* mention Keith Goyden here, however. His jaw-dropping abilities as a researcher were vital to conceiving this project and getting it off the ground. His contributions echo throughout Chapter 4 and into the other chapters as well. I should also acknowledge the Royalty Research Fund at the University of Washington, whose grant also helped launch the project.

I have been extremely fortunate to work with several great people at Routledge. Dave McBride guided the book into the process with grace and warmth. Steve Rutter and Anne Horowitz have been thoughtful, patient, and wise as they guided it to completion. Five anonymous reviewers also deserve mention. They were thorough, insightful, and they made this a much better book than it would have been.

Though they have no category, Adam Joyce and Daryl Maeda are utterly important. In oblique but significant ways they have improved the ideas of this book. They are critical, smart, witty, well-read, and faithful. And, on top of that, they manage to be admirable human beings.

Most important, by far, is my family. My mother, Sue Purcell, has taught me to value learning, but also, by her extraordinary example, to be determined, understanding, patient, generous, and wise. My wife, Elham Kazemi, in addition to being the kind of academic I wish I could be, has an extraordinary light in her heart. I am grateful every day that she allows me to share it. Last but, oh my, not least are Roshann Kazemi Purcell and Neeku Kazemi Purcell. You are quite simply the greatest good fortune of my life.

M.P.
Victrola, Capitol Hill, Seattle
June, 2007

ACKNOWLEDGEMENT OF PERMISSIONS

The author and the publisher are grateful for permission to reproduce the following material:

Cover art: Reproduced from: "Democracy?" (2003). Acrylic, 33 ½" × 41" by Orysia Sinitowich. Her art often reflects the concerns of the past, present, and future of her ancestral homeland, Ukraine. Reproduced with kind permission of the artist.

Figure 4.3 Base map produced with King County's Census Viewer. Reproduced with permission.

Figure 4.4 From Mulady, K. (2004) Remaking South Lake Union: Seattle Is on Fast Track to Build Biotech Hub. *Seattle Post-Intelligencer*. October 20, p. A1. Reproduced with permission.

Figure 4.5 Map produced by CityDesign, City of Seattle. Reproduced with permission.

Figure 4.6 Aerial photo © The Sanborn Map Company Inc. Reproduced with permission.

Figure 4.7 © Washington State Department of Transportation. Reproduced with permission.

Figure 4.8 Map produced by CityDesign, City of Seattle. Reproduced with permission.

Figure 4.9 © King County Water and Land Resources Division. Reproduced with permission.

Figure 4.11 Map created by EPA. Reproduced with permission.

Figure 4.12 Aerial photo © The Sanborn Map Company Inc. Reproduced with permission.

Figure 4.13 Map originally published in Purcell, M. (2001) Metropolitan Political Reorganization and the Political Economy of Urban Growth: The

Case of San Fernando Valley Secession. *Political Geography* 20(5): 101–21. © Elsevier Science Ltd. Reproduced by permission.

Figure 4.14 Map originally published in Purcell, M. (2001) Neighborhood Activism among Homeowners as a Politics of Space. *Professional Geographer* 53(2): 178–94. © Wiley-Blackwell. Reproduced by permission.

Figure 4.15 Map originally published in Purcell, M. (2001) Neighborhood Activism among Homeowners as a Politics of Space. *Professional Geographer* 53(2): 178–94. © Wiley-Blackwell. Reproduced by permission.

Figure 4.18 Aerial photo © The Sanborn Map Company Inc. Reproduced with permission.

INTRODUCTION

In January of 2003, the novelist Arundhati Roy spoke at the World Social Forum in Porto Alegre, Brazil. "Remember this," she said, "we be many and they be few. They need us more than we need them. Another world is not only possible, she is on her way. On a quiet day, I can hear her breathing." This book is an insistent claim that Roy is right. The movements to bring about another world are already in motion. Some are large, others small. Some are intense bursts, others are quietly simmering. They are gathering in places like Seattle, Cancun, Davos, Doha, Goteborg, and Genoa to demand a more democratic and socially just global economy. They are organizing massive street protests all around the world to decry the U.S. invasion of Iraq. They are coming together each year at the World Social Forum to share analysis and strategy. In India, the Save the Narmada movement is resisting the construction of dams along the Narmada River, opposing the displacement of poor and tribal people, and working to preserve ecosystems. In Brazil, the Landless Workers Movement is helping dismantle the country's massively unequal distribution of land. In Chiapas, the Zapatista movement struggles to maintain native peoples' control over their land and resources in the face of neoliberal assaults. In 2006 across the United States a "national day of action for immigration justice" served notice that undocumented immigrants are a political force that will not be silent as the country debates its immigration policy. And it is not only another world that is possible, but another city too. In Los Angeles, low-income, immigrant, and non-white bus riders are demanding, and achieving, affordable, clean, and dignified transportation. In the favelas of Brazil, movements claiming a right to the city have helped bring about a national law that challenges the hegemony of property rights by assigning equal importance to the social use value of urban land. In Vancouver, a remarkable movement among intravenous drug users has helped achieve a safe injection site that greatly reduces their risk of overdose and disease. And as we will see, in Seattle a movement that includes environmentalists, native peoples, neighborhood residents, justice activists, and small businesses has fought to democratize a Superfund cleanup on the city's badly polluted main river. The list goes on.

However, despite the energy, creativity, and power of these movements, they face a daunting challenge. Over the past 30 years or so, the global economy, and cities in particular, have been increasingly "neoliberalized." That is to say social life has become increasingly subjected to the logic of neoliberalism: free markets, competitive relations, and minimal state regulation of capital. The result for cities has been an intensification of the competition among urban areas for capital investment. Economic growth has become the dominant imperative for urban policy and planning. As a result, urban land is seen primarily as property, and maximizing its exchange value is the dominant concern. The property rights of owners greatly outweigh other claims by subordinate groups. Governments expand the assistance they provide to capital interests, even as they back away from social commitments to their citizens. In that context, democratic decision-making is often seen as messy, slow, and inefficient; it is a luxury cities competing desperately for investment cannot afford. The result of neoliberalization, many argue, is that cities are becoming ever more unequal, segregated, unhealthy, and oppressive (e.g. International Network for Urban Research and Action 2003).

A great deal of excellent work in geographical political economy (and beyond) has documented neoliberalization in meticulous detail. We have learned about neoliberalism's origins, its logics, and its corrosive effects on cities and on the global political economy more generally. We have learned how neoliberals have steadily progressed from a fringe faction to the dominant voice in policymaking (Harvey 2005). Their ideas, once marginalized, have become unquestioned assumptions. The research argues that while this agenda has been increasingly successful virtually everywhere on the planet, it has advanced farthest in cities. The neoliberal imagination has become the dominant way to imagine the urban future. I admire and am indebted to that research. It has done the crucial work of understanding and critiquing the processes of neoliberalization. While this work is copious and thorough, it is far from finished; there are still nuances we have yet to grasp. Moreover, neoliberalization is continually adapting and mutating. Our understanding of it must continue to develop as well. But despite the continued relevance of that project, my mission in this book is different. I do not focus so much on an analysis of the problem, but on developing solutions. My goal is not to further document the dominance of neoliberalism and its urban future, but to imagine and promote alternatives to it. There is a danger that as we develop a robust critique of the various injustices of neoliberalization, we will focus only on the doors it is closing. If we highlight instead the contradictions of, and the emerging resistance to, neoliberalization, we can see clearly the countless opportunities it is leaving open (Larner 2000). If they are alert and organized, oppositional movements can take advantage of those opportunities to destabilize existing logics and advocate for alternatives. Moreover, those who oppose neoliberalization must do more than merely point to its contradictions and identify accompanying opportunities. We must also find and make known the resistance that is already taking place. Oppositional movements are all over the urban landscape; they are finding and entering the open doors. We must understand and learn from their experiences, and we must contribute to their success.

In this book, I conceive of alternatives to neoliberalization in terms of *democracy*, as democratic urban futures. One promising way to resist the ongoing project to neoliberalize cities, I argue, is to pursue a counter-project to democratize them. To do so, we must imagine, foster, and publicize *democratic movements* that reject the dominant neoliberal assumptions and pursue more just, more humane, and more socially cooperative urban futures. However, such an approach requires caution. Democracy is a malleable and contested concept. Just what it would mean to "democratize" cities is not at all clear from the word itself. Democracy is not necessarily and naturally opposed to neoliberalism (or capitalism more generally), rather it is only potentially so. Democratization must be consciously constructed in a way that stands against neoliberalization. In order to do that, I develop a set of *democratic attitudes* that explicitly resist neoliberalization and pursue instead the radical democratization of cities. I hope those attitudes can inform, inspire, and support democratic movements as they pursue alternative urban futures.

Such democratic attitudes are best generated and refined by means of a dialogue between theoretical reflection and concrete practice. That is, I do not aspire to produce fully formed democratic principles out of the theoretical ether, nor do I merely report a set of practices that activists are currently pursuing. Rather I aim at a dialogue through which empirical cases inform theoretical understanding, an understanding that can in turn inform democratic practice. Democracy is a much studied and debated concept, and those discussions offer great theoretical wisdom that we should draw on in building successful movements. But any cursory examination of actually existing democratic movements quickly reveals the limitations and oversimplifications of theoretical models of democracy. Actual struggles nearly always involve complex and shifting democratic values and strategies; they are always only imperfect manifestations of democratic principles. So it is difficult and perhaps even counterproductive to construct *a priori* a rigid doctrine for democratic practice and expect urban actors to hew to it.

At the same time, it is critical to engage generalizations, to try to make some sense in general of the complex politics actors are faced with. Theoretical reflection can offer movements helpful signposts in the shifting sands of contemporary urban politics. And it can suggest practices and strategies that might travel, that might be pursued effectively across many struggles. Of course, such a process of making sense through theory always happens within particular grounded contexts. For the purposes of this book, those contexts include the wider historical context of neoliberalization in U.S. cities, the specific manifestation of neoliberalization in Seattle and Los Angeles, and the myriad local specificities that are particular to each case. Any generalizations are suggestive and arise out of the American urban context. But the book does not shun generalizations; it does not refuse to acknowledge generally similar processes (such as neoliberalization) that reappear in a variety of contexts. However, it also takes very seriously the wide variation in how general processes play out. It therefore acknowledges the need to build movements that are significantly tailored to particular political milieus, times, and places. Each movement is the best judge of which democratic strategies will work best in their particular context. However, I argue that movements should make such judgments in light of dialogues: both with other movements in other contexts, and with the wisdom of existing theoretical reflection.

The result of this dialogue is therefore not a democratic model or rubric; it is not a list of strict principles. It is rather a set of democratic *attitudes*. I use that word hoping to evoke its many meanings. In its everyday sense, an attitude is an emotional and mental stance toward the world, a stance that implies some sense of particular values. In its natural-science sense, attitude implies a characteristic, habitual way to respond to a particular stimulus. In terms of physics, an attitude is the physical position a body presents to its environment. Such a meaning evokes Antonio Gramsci's notion of a "war of position," through which a movement aims to increase the influence of their alternative way of perceiving the world. Lastly, more colloquially, having an attitude implies an oppositional, confrontational approach to the world. By "democratic attitudes," then, I mean a set of habitual democratic values that position a movement in opposition to dominant neoliberal values. Those values are "habitual" in the sense that they are a routinized baseline—a first tendency—that need not be worked out anew every time a new political situation arises. But attitudes are suggestive rather than rigid, they can be changed and adapted if further reflection suggests they are not the best way to approach a particular situation. They are, I hope, useful *tools* that movements can employ creatively to their advantage.

To develop these democratic attitudes, I articulate one iteration of a dialogue between theoretical reflection and actual practice. I begin (in Chapters 1, 2, and 3) by offering a theoretical "opening statement:" an analysis of neoliberalization and democracy. Then, as an empirical "response" I present four case studies

(Chapter 4) in which neoliberalization and democracy engage each other. The last chapter reflects on what lessons we can learn from that encounter between theory and practice. The specific chapters proceed as follows. Chapter 1 provides an account of the existing understanding about neoliberalization so that we can better grasp the complex relationship between neoliberalization and democracy. Understanding the nuance of that relationship will help us better imagine what democratic resistance and alternative futures might look like. Chapter 2 offers a careful consideration of current democratic theory to understand which existing democratic ideas are most promising for confronting neoliberalization.

Out of those ideas, Chapter 3 builds a set of democratic attitudes for moving toward a better urban future. It argues that in addition to their particular democratic politics, those attitudes must be fully urban and spatial. Drawing inspiration from Henri Lefebvre's concept of "the right to the city," the chapter contends that building more democratic urban futures is necessarily a spatial project. Chapter 4 then brings that theoretical "opening statement" into dialogue with empirical case studies from Seattle and Los Angeles. The cases help us understand better the possible pitfalls of democratic movements, but they also reveal an astonishing potential. Despite what seem like the worst of times for cities, democratic movements are not bemoaning the many ways they cannot participate. They are instead exploiting the many ways they can. Another world is on her way.

CHAPTER 1

THE "TERROR" OF NEOLIBERALIZATION[1]

THE WASHINGTON COMPETITIVENESS COUNCIL

In the spring of 2001, Boeing announced it was moving its corporate head-quarters from Seattle to Chicago. While much of their manufacturing activities remained in place, the loss of the Boeing headquarters was widely viewed in the State of Washington as a warning: if we don't make the region more attractive to business, more and more companies will follow Boeing's lead. State government acted swiftly. In the summer Democratic Governor Gary Locke created the "Washington Competitiveness Council," an appointed body whose mission was to evaluate current policy and recommend ways to make the state better able to attract capital investment. The 35-member council brought together captains of industry, representatives from state and local government, and two labor leaders. Over several years, it produced an agenda that had two principal components. The first stressed the importance of creating general agreement in the public sphere—a "common sense"—that Washington State must make itself as competitive as possible in the global economy. It stressed the need to "improve public understanding of the importance of a healthy business climate to the future of Washington's economy" (Washington Competitiveness Council 2002, p. 4). The value of local competitiveness, it felt, should be an accepted necessity in policymaking. The second component followed the first by proposing a concrete set of policy recommendations. In order for the state to remain competitive, they found, government should simultaneously get out of the way and actively intervene. It should get out of the way by reducing taxes and fees; "streamlining" environmental, labor, and planning regulations; and creating a customer-friendly ethic in regulatory agencies. It should actively intervene by significantly increasing public spending in several ways: improve transportation, telecommunications, and water infrastructures; fund innovative R&D activities; and restructure public education to produce a more competitive workforce. Without such action, the Council warned, "Washington will fall behind other states and regions that are investing massively in these areas" (Washington Competitiveness Council 2002, p. 19). While the council's recommendations were only advisory, it was extremely successful at turning its recommendations into real policy. Where possible, the Governor issued executive orders to realize the council's agenda. More often, policy change required legislative action, and the legislature was accommodating. The council's first report in 2001 made 99 policy recommendations, and over the course of two sessions the legislature passed 46 bills that advanced the council's agenda (Washington Competitiveness Council 2003, 2004).

Of course these policy changes were favorable, in general, to the profits of the firms represented on the Council. They were proposing policy changes that significantly benefited their bottom line. But the Council's activities were far from covert. It did not meet in smoke-filled back rooms, hidden from public view. It met in open forums at universities, convention centers, and at the state capitol. Its activities and reports were widely publicized. Graft and secret influence were not driving the agenda. Rather, there was broad consensus in the public sphere that government and business should work together to meet the needs of business. That extraordinary argument was the basis of the Council's mission, and it was taken to be self-evident. Far from hiding his relationship to the council, the governor trumpeted it loudly as an important civic undertaking. He was always clear that keeping the state economy competitive was a central responsibility of his office. His official website stressed his accomplishments in this area:

> Gary Locke has led efforts to make Washington a place where businesses want to locate or expand. He's cut more than $1 billion in businesses taxes, worked to expand and promote exports, advanced workforce training, and invested in rural economic development.

The governor made it clear that "we must improve our competitive edge to ensure a healthy business climate" and that "in a rapidly changing global economy, the state must keep pace with changes necessary to keep and improve its competitive edge."

GLOBALIZATION AND NEOLIBERALIZATION

The story of the WCC will come as no surprise to anyone who has followed research in political economy over the past two decades. A great deal of excellent work has documented in meticulous detail how, over the past 35 years or so, the global political economy has been subjected to a massive reorganization. One shift has been a complex rescaling of economic activity and political control, often labeled "globalization." The second shift is tightly tied to the first: a concomitant change in the content of political-economic thinking, what I call "neoliberalization." There has been a massive amount written about globalization, and there is a smaller but still-large body of work on the project of neoliberalization. Therefore, before I discuss the challenges neoliberalism poses to democracy, it is important to be clear about how I understand both elements of this dual shift.

Globalization as Selective "Glocalization"

Over the past three decades, both the global economy and the nation-state have been significantly rescaled. During the postwar era (up to about 1970), political and economic co-ordination were contained to a significant degree at the national scale. The scale of state sovereignty was successfully constructed as fundamentally national, and that scale took on a leading role in organizing both political and economic life (Agnew 1994). National economies tended to interact with each other at arm's length (Dicken 1998). To be sure, non-national scales were also important during this period, but the national scale enjoyed an extended period as the hegemonic scale for co-ordination of both political control and economic activity. The current era has seen a strategic erosion of that hegemony. To a much greater degree than previously, non-national scales are being promoted as alternatives to the national scale. These alternative scales are both supra-national and sub-national, and they affect both economic activity and state control. As a result, many in the literature use the term "glocalization" rather than globalization, in order to more accurately describe this rescaling process (Swyngedouw 1992; Courchene 1995; Robertson 1995).

Both the national scale and its emerging alternatives are assumed here to be socially constructed. That is, they do not exist as fixed entities, each assigned a characteristic set of processes. Instead each scale and the relationships among scales are contested and continually reproduced. Research in geography on the "politics of scale" has argued that scale is always fluid. Scales and scalar relationships can become fixed, but that fixity is never permanent, and it is always realized through a political project (Smith 1993, 1995; Swyngedouw 1992, 1997; MacLeod and Goodwin 1999b; Marston 2000; Brenner 2001). If scales are produced through struggle and conflict, the extent and content of each scale, as well as its relationship to other scales, must be the result of particular groups achieving a configuration that best suits their agenda. Scale, in other words, is best seen not as a neutral container that exists outside politics, but as a *strategy*, as a way to pursue a political agenda. In this light, the political-economic literature has developed an overarching argument that the glocalization of scalar relationships is a conscious strategy to advance the agenda of neoliberalization.

Economic Glocalization

The rise of the transnational corporation and the increasing internationalization of economic production and finance since 1970 have significantly expanded the scale at which capital investment, economic production, and information flows are functionally integrated. Capital flows and economic production are much more likely to extend across national borders than before. That "globalization" of its operations has been an important strategy on the part of capital to achieve two goals. The first is mostly economic: to seek out new areas for investment that could help reverse the declining rate of profit that developed in the industrialized economies due mostly to high wages earned through successful labor organizing. Globalization can be seen, therefore, as a contemporary manifestation of an age-old capitalist strategy: a restless effort to incorporate new territories into the capitalist relation in order to secure new spaces that offer greater returns on investment (Hardt and Negri 2000). The second goal of globalization is more fully political. Under the postwar "Keynesian" policy regime, labor achieved a strong political position: unions had a formal role in macroeconomic policymaking. Labor reproduction and demand stimulation were the main concerns of policymakers. National government tightly regulated capital in a variety of ways. Against the strength of labor, capital's primary bargaining leverage was its mobility, and so transnationalization of investment was a way to give this threat very sharp teeth (Bowles and Gintis 1986, pp. 57–8). Capital has thus "jumped scales" beyond the national as one way to overcome the limitations and contradictions of the predominantly national-scale regime of the Keynesian era (Smith 1993).

The other aspect of economic glocalization is the increasing tendency for economic coordination to be organized at more local and regional scales. Commentators have stressed the growing importance of regional economies as important functional nodes in the world's economic geography (Storper 1997; Scott 1996, 1998). The rising importance of regions is associated with a move toward industrial production that relies on "flexible specialization," whereby very large firms give way to a proliferation of smaller ones (a process known as vertical disintegration) that must coordinate their activities (Piore and Sabel 1984; Christopherson and Storper 1989; Sabel 1994; Saxenian 1994; Hirst and Zeitlin 1989; Scott 1988). As more industries move toward vertical disintegration and flexibility, interfirm cooperation becomes increasingly crucial to economic production. The need for cooperation encourages firms to cluster geographically in a given region and to develop a regional-scale organization that allows (usually urbanized) regions to function as nodes in the global economy (Scott 1996).[2] This economic regionalization is important because it has made the economic fortunes of regions relatively more independent of their national economies. A regional economy is now more able to grow when the national economy is stagnating, or stagnate as the national economy grows. A region's economic networks are no longer just national, but global. That emerging independence of regional economies has intensified inter-region competition for capital investment. No longer are regional economies a function of, and protected by, their national economy. Increasingly they must hustle in the global economy to ensure their economic fortunes.

State Glocalization

The nation-state has also undergone a process of glocalization. As with economic coordination, the national scale has been destabilized as the dominant scale of state power (Jessop 2000). Important state functions and powers are being shifted to other scales, both supranational and sub-national. In some cases, shifts to larger scales have involved merely *inter*national coordination of specific policies whereby each nation-state operates as a sovereign unit within an agreement among nation-states. The UN, NAFTA, ASEAN, and the Kyoto Protocol are examples. In addition, there have been some moves towards truly trans- or supra-national state forms, under which national states partly coalesce into a larger "superstate." The European Union is one of the few examples of an institution in which state sovereignties have the potential to dissolve into a truly supra-national state form (Leitner 1997; Balibar 1999). In both cases, however, both inter- and supra-national state forms have expanded the potential to coordinate policy at scales beyond the national. Of course such international coordination *could* be used to more tightly regulate internationalizing capital. However, it has so far been used much more frequently to aid capital mobility by developing free-trade policies that speed the flow of goods and capital across national borders. Such policies do much to intensify the political leverage associated with capital's mobility.

The converse of state globalization has been a devolution of authority from the national state to its sub-national units (Jessop 1994a; Peck and Tickell 1994; Mayer 1994; Swyngedouw 1996; Ward 2000; MacLeod and Goodwin 1999b; Staeheli *et al.* 1997; Kearns 1995). That process has involved the transfer of functions, obligations, and expectations to local states at various scales, from province/state to municipal. While this is a complex and uneven process, one main outcome has been for the national state to weaken policies that redistributed wealth more evenly across the nation, protecting vulnerable regions from competition. It has instead preferred to stimulate entrepreneurialism by encouraging newly empowered regional authorities to compete among themselves. Erik Swyngedouw's (1996) analysis of the redevelopment of regions in Belgium after mine closures shows how a nexus of new local-state (and quasi-state) institutions were created to assume responsibility for functions such as unemployment, education, economic development, and finance. The goal of the devolution, he argues, was to "produce competitive regional spaces" through institutional state forms that fit more closely the scalar structure of the changing economic geography in the area (1996, p. 1499). According to Swyngedouw, the political project was to shift state policy away from a national-redistributive project and toward a regional-competitive one. With devolution, each region could offer a different regulatory package, and attracting capital became an increasingly important driver of that policy, over and above social need.

As the Belgian example hints, a third shift accompanies this rescaling: a move from formal government to more informal "governance" arrangements. In addition to upscaling and downscaling, there has been a tendency toward outsourcing of state functions to non-state and quasi-state institutions. The state has increasingly privatized and semi-privatized its functions by contracting out services to volunteer organizations, community associations, non-profit corporations, foundations, and private firms, and by developing quasi-public bodies, such as QUANGOs, training and enterprise councils, appointed competitiveness councils (like the WCC), urban development corporations, regional development authorities, and public–private partnerships, to carry out the functions of formal government (Goodwin 1991; Krumholz 1999; Payne and Skelcher 1997; Walzer and York 1998; Watson 1995). That shift has helped increase the "flexibility" of policy-making: very often it is a way to undermine government requirements and protections, because those may cease to apply when a non-governmental agency takes control of a decision. The neoliberal goal here is to move governance out of the routinized channels of the formal state, because that had been the bastion of the Keynesian compromise. The result has been an increasing proliferation of ad hoc and special purpose entities, most of which are quite new. They have missions, rules, and procedures that are not well-established. Such "flexibilization" has created a much more open and unstable set of governing arrangements for urban political actors to navigate. While that increased openness is intended to benefit capital, it often leaves open significant opportunities that subordinate interests can exploit.

As with other processes of rescaling, the process of state rescaling has by no means been total. On the contrary, it has been markedly partial and uneven. The argument is not that the governing and coordination has been diminished; rather governing power has been partly "displaced" from its dominant location in the national-scale state and "replaced" in new institutional and scalar forms, both state and non-state (Jessop 1994b, p. 24). In short, state power has been partly glocalized, and the national-scale state no longer holds as dominant a position as it once did (Brenner 1997; Jones and Keating 1995; Amin 1994; MacLeod and Goodwin 1999b). As the analysis above begins to suggest, rescaling has not been absent of political-economic content. Rather, glocalization has been pursued as a conscious *strategy* to realize a particular agenda: a thoroughgoing neoliberalization of political economies (Swyngedouw 1997). Glocalization has been partial because it has been selective: neoliberals have pursued primarily those scalar shifts that augment the dominance of competitive free-market relations. In order to more fully understand neoliberal globalization, the next section examines that neoliberal agenda in more detail.

Neoliberalization

Those rescaling processes—the increasing threat of capital mobility, the intensification of inter-region and inter-urban competition, and the glocalization of both state regulation and economic production—have all worked in close cooperation with the rise of neoliberalism. The latter is in many ways the reassertion of an old neoclassical economic argument: society functions better under a market logic than any other logic, especially a state-command one. We should give firms and individuals freer reign, neoliberals argue, so that they can rationally maximize their private economic interests in open and competitive markets. In the postwar era, until about 1970, a "Keynesian" economic policy regime instituted strong union power, significant state control over the economy and regulation of capital, and a relatively large welfare state apparatus (Jessop 1993). Even as Keynesianism became dominant, opponents were building an argument for an alternative, a neoliberal ethic in which the state would play a minimal role in the economy and "the invisible hand" of market decisions would determine economic outcomes. The neoliberal argument that the market is far more "efficient" in allocating resources than the state or other institutions, true or not, resonated strongly in the 1970s era of stagflation and economic recession. It has since become the dominant assumption in policymaking.

In the neoliberal imagination, open and competitive markets not only produce the most efficient allocation of resources, but they also stimulate innovation and economic growth. Thus market logics and competitive discipline should be fostered in the economy, and they should even be extended beyond the economy, to institutions like the state, universities, hospitals, schools, and so on. Moreover, because state policies are the primary impediment to competitive markets, the state should "get out of the way" as much as possible. Those values have informed a wide-ranging policy agenda. Neoliberals have worked to reduce international trade restrictions and establish the dominance of a "free trade" ethic to support the globalization of production. They have dismantled the Breton Woods system of fixed exchange rates for currency and allowed money markets to operate much more freely. They have vigorously pursued "flexible" labor markets that are less encumbered by practices like collective bargaining, minimum wages, job security, health insurance, safety laws, and the like. They have pressed for the legal expansion of private property rights to promote the commodification and private ownership of social products such as material goods, land, ideas, and information.[3]

But we should not see neoliberalism only as a concrete economic policy agenda. As Harvey (2005) makes clear, the rise of neoliberalism constituted a successful ideological project to make competitive-market ideologies hegemonic. Wendy Larner (2000) and Henry Giroux (2004) have both emphasized the importance of understanding neoliberalism not just as a set of concrete policies, but as an ideology, a form of governmentality, and, as Giroux terms it, a "public pedagogy." The hegemony of neoliberal ideas is important, because creating a dominant neoliberal "common-sense" helps establish unquestioned assumptions that make it very difficult to imagine, let alone achieve, alternative projects (Keil 2002). For example, in laying out their policy agenda, the Washington Competitiveness Council made very clear the need to continually "improve public understanding of the importance of a healthy business climate to the future of Washington's economy" (Washington Competitiveness Council 2002, p. 4). That exhortation suggests two things: (1) proponents of neoliberalism understand very well the importance of Giroux's "public pedagogy," and (2) the dominance of neoliberal common-sense is not yet total. The WCC was concerned that not everyone in Washington State had accepted the importance of their project (although they were confident education could help them "understand"). While the project of realizing neoliberalism's ideals is important, nowhere it is complete or permanent. Rather what we are seeing is an ongoing struggle to continually re-establish neoliberalism as the hegemonic logic for political argument (Gramsci 1971; Jessop 1997a). The more dominant this logic is, the easier it is to advance the neoliberal policy agenda. Much headway has been made in realizing that agenda, but we should not expect to see any pure manifestations of a neoliberal political economy. Rather, if we examine the concrete policy agenda of neoliberalism, what we see is a distinctly fitful advance and uneven geography: some places labor under a more nearly neoliberal regime, in other places neoliberalization is much less far along. It is that sense that I use the term neoliberalization, as an ongoing but never completed project to neoliberalize urban political economies (Tickell and Peck 2003).

Adding to that complexity is a nuance within the policy ensemble that has emerged under neoliberalization. While neoliberal doctrine propounds a minimal state and maximal markets, the discussion above suggests a more complicated agenda. "Neoliberalism," Hardt and Negri (2004, p. 280) write, "is not really a regime of unregulated capital but rather a form of state regulation that best facilitates the global movements and profit of capital." The state is not merely stepping back, rather it is dismantling its assistance for labor and the poor and increasing its assistance for capital (Moody 1997). Peck and Tickell have labeled this the "roll-back" of economic regulation and welfare spending and the "roll-out" of pro-market policies (Peck and Tickell 2002a). Brenner and Theodore prefer to call them the "moment of destruction" and the "moment of creation" (Brenner and Theodore 2002). Because I want to highlight specifically how neoliberal policy favors capital by both helping and getting out of the way, I use the terms "laissez-faire" to describe policies that lower barriers for capital and "aidez-faire" to describe policies that actively help capital to accumulate.[4] In their account, Peck and Tickell periodize these two aspects of neoliberalism, arguing that an earlier phase of roll-back in the 1980s was followed in the 1990s by a roll-out of policies to increase state support for capital. While this historical big picture is important to understand, for a given place it is most useful to think of these two aspects as occurring together in a complex mixture of both laissez- and aidez-faire. Neoliberalization is a project to both destroy impediments and create assistance to capital wherever and whenever possible.

Conceived of in Gramscian terms, as a hegemonic project, neoliberalization is always only imperfectly realized. It does not exist in its entirety because (1) it must articulate with and accommodate existing policies, habits, and assumptions, such as those remaining from the Keynesian era (Brenner 2005), (2) it produces its own contradictions and legitimacy problems, and (3) as a result, it is always resisted. That does not mean that neoliberal logics are not dominant—they are. But it does mean their dominance is never total. The advance of both laissez-faire and aidez-faire are partial, uneven, and highly context-specific. Neoliberals have to be opportunistic and adaptive in pressing their agenda. They must slot in with existing policy, regulate internal contradictions, and overcome resistance. That non-monolithic character is important because it helps us realize that neoliberalism is not invincible, it is merely the current hegemon in a long line of hegemons. Counter-projects are possible; indeed they are inevitable. But it is also important because it helps us understand better the actually existing policy frameworks that neoliberalization has produced (Larner 2000).

Laissez-faire

So neoliberalization is not so much state policy getting out of the way as it is transforming policy to better meet the needs of capital. However, a central part of this agenda is the laissez-faire ideal whereby state regulation of economic activity is minimized. Perhaps the most well-known example is the reductions of tariffs and other regulations to promote international free trade, a policy ensemble that was one target of the Seattle protests. Within nation-states, laissez-faire minimizes environmental regulations like the U.S. Clean Air Act that impose "external" restrictions on economic decisions, or the Endangered Species Act that often limit what a property owner can do with his or her land. Government regulations that codify collective bargaining and other workplace rights (National Labor Relations Act), or promote safe working conditions (Occupational Safety and Health Administration (OSHA)) are undermined, empowering capital at the expense of labor. Similarly under attack are programs like unemployment insurance, since these "distort" the labor market because they mitigate the true cost of being fired and therefore make it harder for capital to discipline labor. Neoliberalization also reduces regulations for spatial planning, like Washington State's robust growth-management limits on land development, or the national "spatial Keynesian" policies that favor declining regions in order to ensure a more equal distribution of wealth (Martin and Sunley 1997; Brenner 2005). Laissez-faire also targets taxes paid by firms for reduction or elimination. And it pushes localities to reduce tax burdens in general so they remain competitive with other localities. Neoliberalization undermines the regulation of housing markets, seeking to eliminate or streamline policies like rent controls, building codes and permits, and affordable housing requirements. It also aims to minimize a host of similar regulatory structures, such as those associated with insurance, business licensing, and liability.

The shift to governance, whereby government functions are outsourced to the non-government sector, generally helps advance the goal of deregulation. Because it produces a more "flexible" governing climate, it undermines the potential for stable statutory regulation that the state can provide. Often state requirements (e.g. when a city government has adopted a living wage ordinance, or agreed to abide by the Kyoto Protocol) cease to apply when government functions are outsourced to quasi-public or voluntary organizations. Furthermore, as we saw above, the devolution and downscaling of government functions from the national scale to local and regional agencies increases competition between localities. Whereas localities in a given nation-state were formerly all operating under the same national regulatory structure in a sphere like environmental management, for example, with devolution each locality can now regulate differently, and so localities can compete for capital by reducing their regulatory requirements. As Harvey (2005, p. 87) puts it,

> competition between territories (states, regions, or cities) as to who had the best model for economic development or the best business climate was relatively insignificant in the 1950s and 1960s. Competition of this sort heightened in the more fluid and open systems of trading relations established after 1970.

Such "entrepreneurial" urban governance means there is a very real danger that such inter-local competition is a "race to the bottom" in terms of the social controls (like democratic decision-making) a locality can place on capital (Harvey 1989). In this way, glocalization and neoliberalization together produce competitive discipline that helps dampen the state's ability to regulate capital.

Welfare Retrenchment

Neoliberalization doesn't just reduce particular forms of state regulation, it also reduces state spending in some areas. Perhaps the most important example is that of welfare spending. The robust welfare states that grew out of the Keynesian era have been progressively "downsized" over the past 25 years. A long string of government retreat from programs such as direct aid to families, unemployment insurance, social security, child care, and health care has paraded across the political stage in this era (Staeheli *et al.* 1997). Such welfare programs are disparaged by neoliberals because they are considered uncompetitive. We should invest in productive economic enterprise, they argue, not unproductive social handouts. Workers with access to unemployment insurance or direct government payments will lack the incentive to participate energetically in the labor market. If we remove such security, the argument goes, workers will be highly motivated to succeed in their jobs. Moreover, as we saw above, such safety nets reduce capital's power because they make the threat of termination less worrisome for workers. In general, the neoliberal argument is that we should meet social needs not through the bureaucratic state but through the efficient allocation mechanism of the market. As a result, we see the current push to move people off welfare and into work, or a similar drive to privatize social security by allowing participants to invest their money in private funds. While such welfare retrenchment does not always directly benefit capital like regulatory relaxation does, it does significantly reduce an important commitment to public spending and so makes it more feasible to reduce the tax burden on capital.

Perhaps more than any other area, welfare retrenchment has been subject to state rescaling through devolution (Hoggett 1987). National states have increasingly offloaded responsibility for welfare provision onto local governments. Local governments, in turn, have outsourced a range of responsibilities to non-profit, volunteer, and private organizations. However, this offloading of responsibility has not been accompanied by sufficient state resources to meet the new demands. In fact, national states have progressively dismantled their traditional fiscal support for localities. During the Keynesian era, national governments allocated resources to local governments on the basis of need, social entitlements, and automatic stabilizers (Peck and Tickell 2002a, p. 395). That system helped ensure localities had enough revenue to cover the costs that local need generated. The neoliberal model distributes such funding competitively: central governments are now less inclined to consider need and prefer to invest in localities that seem to promise a return on investment in terms of local economic growth.[5] Cities with promising economic profiles are now more likely to secure central funding than cities with greater social need. That worsens uneven spatial development as positive feedbacks develop for richer places (growth attracts funding to stimulate growth) and poorer places are burdened with negative feedbacks. For the latter, an acute "fiscal squeeze" develops because they have increasing social need and decreasing revenue to meet it. That squeeze intensifies the need for these places to generate revenue locally. But raising local taxes for welfare spending dampens competitiveness, and so local governments are pushed to entrepreneurial strategies that attract new investment that can increase tax increments (Harvey 1989).[6] So the inter-local competition that cities are increasingly exposed to is intensified by their rising need to stimulate economic development that can generate badly needed public revenue to meet their increased service burden. In short, welfare retrenchment and devolution have been an important incentive for local policy to be more competitive and entrepreneurial.

Another form of retrenchment involves rolling back state protections that aid particular industries. States cut aid to favored or struggling national industries (e.g. steel or cotton), thereby forcing their firms to compete globally. In places where such industries are or were an important part of the local economy (like in the rust belt), their decline or wholesale relocation has further pushed local governments to be more entrepreneurial. The spatial Keynesianism of the old order would have mobilized national-government revenue to mitigate such economic hardship and preserve a measure of inter-local equality. The neoliberal order, on the contrary, tends to leave such places to their own devices, forcing them to invent new entrepreneurial strategies to reinvigorate their competitiveness. In addition to those pressures, one additional assault on state spending concerns public-sector employment. The "lean, mean" government of the neoliberal imagination encourages continual reduction of the number of state employees, and vigorous resistance to their strong unions. The large payrolls of "big-government" run afoul of the neoliberal emphasis on efficiency. That labor power, they argue, would be more productively applied in the competitive private sector.

Aidez-faire

Welfare retrenchment fits well with the neoliberal ideal of a minimal state that gets out of the way of the free market. But in fact, neoliberalization has involved both the retrenchment of state spending and its augmentation. Seen as a whole, the agenda of neoliberalization is to reduce state spending that does not benefit capital in order to free up revenue for spending that does. More specifically, capital needs the state to pay for things it cannot or does not want to provide for itself. For example, massive public investments in infrastructure projects that are necessary for economic growth are key to the agenda of neoliberalization. For infrastructure that has traditionally been the responsibility of the state, like transportation, there has simply been an intensification of pressure to maintain efficient flow of people and goods. The example of the Washington Competitiveness Council is again instructive: their "most imperative recommendation" for increasing local competitiveness was to "fix our transportation problem" (Washington Competitiveness Council 2002, p. 8). Of course such spending requires revenue, and the WCC imagined that revenue would be publicly provided. However, because their agenda is also to reduce existing taxes on business, they called for "alternative financing mechanisms" to raise revenue from sources other than businesses (Washington Competitiveness Council 2002, p. 10). The neoliberal hope, or perhaps more accurately, faith, is that some of that new money will come from savings realized by welfare retrenchment, but most will come from new revenue produced by economic growth. That contradiction between reduced revenue and the demand for increased spending is endemic to the neoliberal agenda. It requires creative new ways to raise needed public revenue in a political economy that demands lean, mean government. That is one reason why inter-local competition is so crucial: governments feel they must find the revenue for infrastructure somehow, even at the cost of reducing all other spending, because without effective infrastructure they won't be able to compete for investment and the local economy will stagnate.

Other traditional state spending responsibilities have been similarly intensified, but also reimagined. Education spending is regularly a priority for the neoliberal agenda (Gordon and Whitty 1997). But the goal here is not to reinvigorate liberal education to produce critical citizens. Rather it is to refocus educational spending to produce a varied pool of skilled "human capital" that can enhance a place's competitiveness. The idea is to produce both professionals and skilled manufacturing and service workers that, when paired with lower wages relative to other such places, together create an attractive package of labor for capital. Places like Bangalore, Thomas Friedman's great inspiration, have pursued such a strategy to attract significant investment in high-tech production and customer service (Friedman 2005).

Neoliberalism also encourages states to spend beyond traditional areas, pushing them to take on new responsibilities. For example, local and state governments are increasingly becoming principal investors in urban development projects. Rather than merely regulating land development, the state has become much more actively involved in providing investment capital to stimulate development. Here public–private partnerships are commonly used, whereby a state agency teams with a private firm to develop a property. States have always funded the development of public works projects, but they are increasingly providing capital to jumpstart the development of property for sale on the private market. The terms of each deal vary of course, but often the state provides funds simply to get development going, and the private firm receives the bulk of the profit. Even where the profit is shared, state financing reduces the cost and risk firms must assume to develop urban land. Especially important here have been the growing number of "mega-projects" to revitalize and draw attention to a given city (Altshuler and Luberoff 2003; Flyvbjerg *et al.* 2003; Olds 1998). While these are sometimes purely public works projects, increasingly they take the form of public–private partnerships to develop urban land for the private market (Swyngedouw *et al.* 2002).

Another form of state support is technology transfer whereby states hand over technology generated by the public sector to private firms. Organizations like Senator Robert Byrd's National Technology Transfer Center "strengthens the competitiveness of U.S. industry by providing access to federally funded research" (National Technology Transfer Center 2006). The goal of the program is to "commercialize" the technological innovations produced by federal research, essentially to give publicly generated knowledge to capital interests. The public bears the costs of research and development, and private firms realize the profit. That model also operates at smaller scales, as when localities use respected state research universities to attract high-tech industry to places like Silicon Valley, the Raleigh-Durham "research triangle," Austin, Texas, and Seattle, Washington. More directly, states will also "incubate" particular industries they feel are competitive because they are at the leading edge of technological innovation. Software, internet technologies, and biotechnology and other forms of medical research are prominent examples. Here local states will use a combination of laissez-faire (tax breaks, zoning flexibility, etc.) and aidez-faire (subsidies for fledgling firms, technology transfer (especially in biotech), development investment, etc.).

Aidez-faire spending affects social policy as well. In housing, as the state backs away from regulations like rent control and Keynesian spending like public housing, it "rolls out" new forms of spending like housing vouchers and certificates. Those allow property owners to receive the "fair market value" in renting a unit, because the state pays what the tenant cannot. The state increasingly does not develop and own housing, nor does it regulate housing prices; it instead subsidizes low-income housing consumers so they can behave just like other consumers, and the housing market can function as close to the neoliberal ideal as possible. As with other forms of neoliberal restructuring, responsibility for administering such subsidies is increasingly being devolved to the local and regional scale, and the amounts of their payments are pegged to local housing markets.

In the labor market, Keynesian "welfare" policies of direct aid to workers are being replaced by "workfare" systems designed to get workers off the government dole and into the labor market (Peck 2001). Recipients get job training and undergo "work readiness" programs in which they learn literacy, numeracy, and interpersonal skills deemed appropriate for interviews. Ideally, this sort of training is contracted out to private firms. Recipients also endure "experience building" in which they work for little or nothing in order to get more experience in a particular job. Even when they move into the regular labor market, rarely do the available jobs pay much above minimum wage or have any long-term security. However, the neoliberal goal is achieved: moved from public assistance into the labor market, they have been commodified as laborers. That workfare imperative is apparent in education as well, as private technical and vocational schools (ITT Technical Institute, University of Phoenix, etc.) proliferate, and public education at all levels increasingly moves away from the liberal ideal of nurturing critical citizens and toward the neoliberal ideal of training ready workers (Washburn 2005; Geiger 2004; Aronowitz 2001). As with housing, workfare policies are increasingly administered not at the national scale, but by provincial- and county-level agencies. As such policy is devolved and each locality can pursue a different set of policy options, the disciplining forces of inter-local competition and the fiscal squeeze rear their heads.

Aidez-faire need not involve active public spending. It can instead involve the adoption of particular values that benefit capital. For example, competitive discipline forces governments to place the highest priority on the exchange value of space rather than on its use value for people who inhabit it. Governments are not just relaxing planning restrictions so capital is freer to operate; they feel they must actively favor maximizing property values over maximizing use values. The notion of "the highest and best use" is a real-estate term that imagines the true calling of a parcel of land to be the maximum market value it can achieve. Such a value-laden term shows clearly how neoliberalism is an ideological and pedagogical project as well as policy one. In the neoliberal era, highest and best use is increasingly becoming the governing doctrine in spatial planning. It is an enormous help to property owners and developers, and it is a heavy burden for movements concerned with use value. A park, for example, might be the best way to meet the needs of neighborhood inhabitants, but it rarely meets the standard of highest and best use.

And along these same lines, governments are rolling out a host of other new policies that both actively assist capital accumulation and promote interplace competition. Monetarist policies, for example, work to control inflation (for capital) rather than unemployment (for labor). Governments create "new institutional relays" (like the Washington Competitiveness Council) that coordinate and mobilize more competitive policies (Brenner and Theodore 2002, p. 22). They create enterprise zones to attract and work with specific forms of capital (Wilson 2004). They create and disseminate powerful narratives about the importance of competitiveness, the possibility of decline and stagnation, and the absolute necessity of economic growth. And they accept the logic of "fast policy transfer," in which local governments adopt without much modification policies developed in other places to improve competitiveness (Peck and Tickell 2002b). That last trend helps neoliberalization spread more quickly from place to place. In short, neoliberal governments don't just leave capital alone, they also help it directly to accumulate and control wealth.

Disciplining

In addition to direct assistance to capital, the state also provides a more indirect benefit by actively pursuing a wider social stability. To a large degree, the state is responsible for managing the inequality and political tensions that capitalist economies tend to produce (Jessop 1990). In the Keynesian era, stability was achieved by including labor in decision-making and mitigating inequality through redistribution. Neoliberalization has largely dismantled those structures, and so it must seek other means. One has been to vigorously discipline those who do not play the neoliberal game or play it poorly. In a broad sense, the competitiveness ethic of neoliberalism is a pervasive way to discipline deviance. Localities who wish to pursue alternatives to entrepreneurial public policy, for example, are often punished with economic stagnation or capital disinvestment. Even if such a dynamic does not always materialize, the dominance of the competitiveness ethic induces in governments a considerable fear of straying from the logic of growth above all else. More concretely, disciplining can target those disadvantaged by an increasingly market-oriented economy (Peck and Tickell 2002a). The workfare policies mentioned above, for example, discipline employees by withdrawing direct aid and so intensifying the costs of not participating actively in the labor market.

We also see more naked disciplining on everyday behavior, as with the rise of "zero-tolerance" policing policies that crack down on even the pettiest crime in an effort to deter disorder and make public space safer for the middle classes (Kelling and Wilson 1982; Smith 1996). Related are increasingly restrictive ordinances to manage and contain the homeless. One disciplining strategy has been to bring an ethic of "personal responsibility" to homeless services. For example, a recently constructed homeless services center in Seattle offers lunches, showers, toilets, internet access, and phone service. However, those using the center for more than ten days are required to sign a contract outlining their plan for entering the housing and labor markets and "transitioning" out of homelessness. The director of the non-profit corporation that the city contracts to run the center put it this way, "If you're just coming in to use the bathroom, we're ultimately going to grow impatient … We will start ratcheting up the pressure to develop a transition plan. If they do not, we're going to assume our resources could be more profitably applied to other people" (Chan 2006, p. B2). Another anti-homeless strategy has been to criminalize everyday acts of survival (sitting on sidewalks, sleeping on benches, urinating in alleys) so that zero-tolerance policing has the power to remove the homeless from places they are not desired and contain them in other, more marginal places (Davis 1990; Mitchell 2003; Macleod 2002). The "civility laws" advocated by Seattle City Attorney Mark Sidran are a prominent example (England 2006). That more punitive option is part of a wider project to create an ever-larger "prison-industrial complex" to house those who do not participate consistently in the low-wage labor market (Gilmore 2006).

One area in which neoliberal disciplining encounters important contradictions is that of migration. Trade liberalization has produced significant economic upheaval in rural areas of countries in the global South as foreign capital industrializes agricultural production and shifts it to an export orientation (e.g. Mexico and Central America). Migrants from those places often move to cities within their nation-state, but they also migrate to countries in the global North (e.g. the United States). Neoliberalization unequivocally advocates freedom of movement for goods and capital; it is less sure about the movement of labor. On the one hand, both documented and undocumented immigrants are a powerful disciplinary tool in the host labor market. Their citizenship status and racialized devaluation drives their wages down and creates great "flexibility" for employers in certain sectors. On the other hand, powerful nationalist and nativist sentiment has forced national governments to more strictly regulate migration (Purcell and Nevins 2005). Frequently both nativism and neoliberalism are most at home in the same political party (e.g. U.S. Republicans), and that party must balance the labor needs of capital with widespread citizen sentiment against immigrants. The result is usually an ebb and flow of migration enforcement, a combination of nationalist protectionism and selective and tentative liberalization. Those tensions were clearly on display in the U.S. in the 2007 debates about immigration reform.

The contradictions associated with migration are not unique. The neoliberal agenda continually breeds contradiction and encounters resistance. Nevertheless, the broad-sweep agenda is clear: to increase the control of capital over material life. In terms of political decision-making, that agenda can be seen as an attempt to increase the control of capital and decrease the power of "the people," however defined.[7] For that reason, many have raised the concern that neoliberalism (and capitalism more generally) poses a fundamental threat to democracy (e.g. Norgaard 1994; Hardt and Negri 2004; Dryzek 1996; Bowles and Gintis 1986). In the next section I explore this claim, and I analyze the specific relationship between neoliberalization and democracy in cities.

NEOLIBERALIZATION AND DEMOCRACY

Those concerned that neoliberal globalization is undermining democracy often present democracy in a common-sense way, without extensively elaborating the debates and struggles over the concept (e.g. Harvey 2005). It is therefore important to be clear that "democracy" has many interpretations. One's sense of the relationship between neoliberalization and democracy will be deeply influenced by which interpretation one adopts. The dominant form, liberal democracy, is joined by a host of alternatives, such as participatory, deliberative, direct, pluralist, and radical democracy. Liberal democracy is in many ways quite compatible with neoliberal globalization. I develop in the next chapter how neoliberals routinely advocate the spread of liberal democracy as part of their vision. They do so, at least in part, in order to compensate for a central problem in neoliberalization: it tends to produce considerable democratic deficits. It therefore struggles with problems of democratic legitimacy, and more generally, political legitimacy.

One of those democratic deficits is historical. Especially in the early years of neoliberalism's rise (but also later), there was no shortage of cases in which authoritarian government was paired effectively with neoliberal economic policy. Examples include Chile, Iran, China, South Korea, Taiwan, and Singapore (Harvey 2005, p. 120). Such a pairing tars neoliberalization with very negative associations from the point of view of democracy. Here Augusto Pinochet is a very embarrassing poster child. And the examples of anti-democratic means toward neoliberalization are not at all confined to history. It was fairly clear that the United States had at least secretly encouraged if not engineered the recent aborted coup against Venezuelan president Hugo Chavez, one of the most vocal opponents of neoliberalism (Forero 2002). Beyond those associations, a more general deficit concerns the process of deregulation. Since in a liberal democracy the state is the principal institution through which the people are empowered, neoliberalization's drive toward deregulation and outsourcing means that democratically elected governments are less able to manage the economy, and "the people" of liberal democracy are therefore ceding power to markets and the firms that hold power in those markets (Larner 2000). While there are of course problems with the naked claim that the state in liberal democracies truly represents "the people," liberal-democratic governments are certainly far more democratically accountable to the people than are corporations. So, to the extent that neoliberalization succeeds in its agenda to augment the power of corporations vis-à-vis the state (Hardt and Negri 2004), and insofar as liberal-democratic states are the principal representative of the mass of people, neoliberalization produces a democratic deficit because it transfers power from democratic citizens to corporations.

Moreover, the increasing globalization of governance has produced more and more institutions that are not governed through democratic representation, such as the WTO, IMF, and the World Bank (Dryzek 1996, p. 6; Wallach and Sforza 1999; Jawara and Kwa 2003). Decision-makers in such bodies are usually appointed by a national-scale executive and so are not tied in any sort of direct or accountable way to electorates. Moreover, in those institutions the representatives of rich states often successfully use specific procedures to dominate and exclude the representatives of poorer countries (Khor 1999; see also Wainwright 2007). At a more local scale, the same sort of failure of democratic representation occurs. For example, in cities special districts are created to build infrastructure projects in the hopes of keeping the local economy competitive. They are usually governed by boards that are appointed by local-government executives. Even when such projects are unpopular boondoggles, the elected officials who appoint the boards are rarely punished at the polls. Voters have too many other issues to consider. Just before his election as Mayor of Seattle, Greg Nickels was a board member of (and apologist for) Sound Transit, a special agency widely vilified for being inefficient, wasteful, and unaccountable. He was elected mayor anyway.

Other perspectives offer a different set of concerns about neoliberalization's impact on democracy. Advocates of participatory and direct democracy, for example, see political representation as by definition undemocratic. They value more local polities and decry supra-national state forms, even relatively representative ones like the EU, because decision-makers are so far removed from democratic citizens. Social democrats, for their part, condemn the erosion of state policies like social insurance and economic redistribution and the associated growth of material inequality at all scales (Hirst and Thompson 1995). They mourn the neoliberal attack on the more socialized political-economies of the Keynesian era because they were able to ensure greater material equality, and social democrats believe such equality is necessary for a robust democracy. The material inequality that neoliberalization produces means "the very social and material basis for greater political equality—central to the very idea of democracy—has been undermined in many countries" (Gill 1996, p. 215). Evidence of that inequality is provided by longitudinal studies that show it increasing on virtually every index (Task Force on Inequality and American Democracy 2004; Dumenil and Levy 2004; Pollin 2003; Fainstein 2001; Harvey 2005).

Even for liberal democrats, such inequality poses problems. The liberal-democratic tradition posits a strong division between public and private, and generally imagines the economy to be part of the private sphere. That logic allows liberal democracies to countenance a strong measure of social inequality even as they claim the formal political equality necessary for democracy. But that maneuver, as one can imagine, doesn't eliminate the problem. The contradiction (or at least tension) between social inequality and political equality has always posed problems for liberal-democratic thought and practice. The greater the social inequality, the more unstable is the claim that all citizens are equally valued and carry an equal voice (Task Force on Inequality and American Democracy 2004). Since such formal political equality is a minimum requirement for all notions of democracy, increasing social inequality destabilizes liberal democracy's claim to political legitimacy. Neoliberalization has therefore worked to exacerbate a democratic deficit that has long been a thorn in liberal democracy's side.

For other kinds of democrats, even more acute deficits arise. For deliberative democrats like Dryzek (1996), neoliberalism promotes consumer identities over citizen identities (see also Larner 1997; Miller 2007). Neoliberalism values individuals who myopically pursue their material self-interest in the marketplace, not citizens who cultivate their civic virtue in the public square. Deliberative democrat Walter Parker (2003, p. 2) reminds us that the ancient Greek term for such self-interested individuals was "idiots." Parker contrasts "idiots" with "citizens," whose actions are motivated instead by the common good. A democracy, Parker insists, cannot thrive unless it can transform idiots into citizens, but neoliberalization works in exactly the opposite direction. Participatory democrats, for their part, share deliberative democrats' concern with idiots. In addition, they also believe neoliberalization is undermining democracy more generally: "deregulation, privatization, reduction of social services, and curtailments of state spending have been the watchwords, rather than participation, greater responsiveness, more creative and effective forms of democratic state intervention" (Fung and Wright 2003, p. 4) The way to resist neoliberalization, participatory democrats contend, is to democratize decisions "by setting up participatory instruments of nonelite empowerment and public accountability" both inside and outside the state (Nylen 2003, p. 120). Here the "participatory budgeting" experiment in Brazil has been one important model for critical reflection (Baiocchi 2003).

Many in urban studies have focused on the anti-democratic effects neoliberalization has had on urban governance (Brenner 1999; Jessop 1997a; MacLeod and Goodwin 1999a). That body of research outlines how for cities neoliberalization narrows the options open to decision-makers whether they are elected or not (Dryzek 1996). Because of the disciplining force of the perceived need to remain globally competitive, cities are pushed away from redistributive social policies and any other option seen as a threat to economic growth. They are pushed toward entrepreneurial policies by the disciplinary forces I outline above. Cities cannot afford to preserve democratic public spaces for citizens (plazas, parks, squares) when they can develop privatized, revenue-producing spaces for "idiots" (malls, offices, condominiums). They cannot spend their meager revenue on services for disadvantaged neighborhoods because that promises little short-term return to the local economy. Affordable housing regularly loses out to gentrification (Smith 1996). And so no matter how cities organize their decision-making structure, and no matter who is making the decisions, the disciplinary structures of neoliberalism ensure (or make it seem, at least) that cities don't have much choice in making public policy. The result is what Byron Miller (2007, p. 235) calls "limited-capacity urban governance." While capitalism has always limited choices in this way (Gintis 1972), neoliberalization, Miller argues, has intensified those constraints. Virtually all versions of democracy require that democratic citizens be able to choose among many real and viable options. Democracy is thus increasingly squeezed into irrelevance by the competitive discipline of neoliberalization.

In addition to that squeezing, urban decision-making structures are becoming less open to democratic debate. Because they must compete, cities dare not delay in making decisions; they must respond quickly to the global market. Truly democratic decision-making tends to involve political wrangling and vigorous debate; it values transparent disclosure and openness to considering new options. It takes time. It is therefore seen as slow, messy, inefficient, and not likely to produce the kind of bold entrepreneurial decisions that attract and keep capital (Miller 2007). So urban governing institutions are being increasingly "streamlined" so they can foreclose lengthy debate and more quickly respond to market opportunities. They have become less transparent and less accountable to the public (Hoggett 1987). Oligarchic institutions like public–private partnerships, appointed councils, and quasi-public agencies are increasingly making decisions that were formerly made by officials directly elected by the public (Goodwin and Painter 1996; Krumholz 1999; Payne and Skelcher 1997; Walzer and York 1998). Enterprise zones and mega-projects, for example, are creating special appointed bodies that govern only that particular project. And they are proliferating (Swyngedouw et al. 2002). In addition, the logic of "fast policy transfer" dictates that urban governments adopt ready-made policy ensembles developed in other places rather than engage the city's public in generating policy through democratic debate. Citizens and their representatives are increasingly replaced in decision-making by panels of business leaders and economic experts who are perceived to know how best to respond to the competitive global market. Non-profit, volunteer, and community organizations are replacing governments as primary service providers. In short, the decisions that shape the city are increasingly being transferred out of the control of the state and its citizens (Brownhill et al. 1996; Keating 1991; Peck 1998; Tickell and Peck 2003; Ward 2000).

On top of these democratic deficits, neoliberalism's basic values create more general problems with social cohesion. Neoliberals adopt the classical liberal emphasis on individual self-reliance. Famously, Margaret Thatcher once went so far as to say "there is no such thing as society. There are individual men and women, and there are families" (Thatcher 1987). Such naked individualism and atomism has always been problematic for social cohesion: if we are merely a collection of individuals, what makes us a cohesive community? In the absence of a strong sense of a cohesive society, it is all the more difficult to constitute legitimate collective authority. But even a neoliberal regime requires legitimate collective authority, at the very least to enforce contracts, protect property rights, and prevent crime, and at the most, as we have seen, to actively intervene on behalf of capital. So, if we add the more general problem of social cohesion to the democratic deficits discussed above, neoliberalization, by its normal operation, creates important problems of democratic legitimacy. Although of course the degree to which democratic legitimacy poses a problem for neoliberalism will vary from place to place depending on a range of contextual factors, in the big picture neoliberalization cannot proceed without actively managing the social and political instability it generates.

Those legitimacy problems make it necessary to develop political mechanisms that can manage them (Tickell and Peck 2003). Such "regulation" of political instability is entirely typical of capitalism more generally (Jessop 1997b; Brenner and Glick 1991; Aglietta 1979; Lipietz 1992), and neoliberalism's ideological emphasis on the market intensifies the need for political regulation (Tickell and Peck 2003; Peck and Tickell 2002a; Jessop 2002). Increasingly, neoliberal policies have been responding explicitly to their democratic deficits (rather than just wishing them away) by assiduously working to appear more democratic. As I argue below, there is a very strong ideological project to associate neoliberalization with liberal democracy. In addition, deliberative democratic practices, which have become increasingly prevalent in planning recently, are being adopted in many places by neoliberals to create a visible process whose rules are democratic enough to legitimate decisions but still ensure a controlling interest for capital (McGuirk 2001; Flyvbjerg 1998a). The tradition of participatory decision-making in cities has also been subjected to at least partial absorption. After the awful experiences of urban renewal in the 1950s and 1960s, many federal laws mandated community or public involvement in decisions (Environmental Protection Agency (EPA), Endangered Species Act (ESA), Intermodal Surface Transportation Efficiency Act (ISTEA), HOPE VI, etc.). But the rules are usually quite vague. There is significant room to construct a process that ensures decisions fall only within a very narrow purview but are nevertheless seen as democratically legitimate because they involve the public in some way. Either the public's participation can be constrained, as for example in the all-too-common "public hearing" with audience microphones. And even when the public is involved meaningfully, their decisions can still be contained within neoliberal bounds by the disciplining logics I discuss above.

Other strategies to associate neoliberalization with democracy take their cue from neoliberalism itself. Neoliberals play up the undemocratic (and illiberal) characteristics of Keynesian bureaucracy. They claim their way offers a more democratic alternative. And this argument is not merely smoke and mirrors. Keynesianism did suffer, as David Harvey (2005, p. 42) notes, from "intense restrictions on individual possibilities and personal behaviors by state-mandated ... controls." Neoliberals counterpose their values of "individualism, freedom, and liberty" with "the stifling bureaucratic ineptitude of the state apparatus" (Harvey 2005, pp. 56–7). The neoliberal tendency toward state retrenchment and devolution, in this light, is sold as a democratization of decision-making (Mayer 2007). The old national-scale Keynesian policymaking, neoliberals claim, was a top-down imposition on local places of the theories of far-away bureaucrats. Restricting that activist state or at least devolving its authority to the local scale allows places more power to shape decisions to their particular context (this claim is chronicled by Swyngedouw *et al.* 2002, p. 226). During the 1970s the Nixon administration sold the federal retreat from community development, and the move toward block grants, as a way to allow localities to judge for themselves how money should be spent (Green and Haines 2001; Rubin and Rubin 2007). The current fashion to create more "collaborative" decision-making processes, similarly, includes more non-state "stakeholders" in decisions that were formerly made mostly within the state. While these shifts can and do open democratic opportunities, nevertheless if the stakeholders are carefully selected, and they are effectively disciplined by the competitiveness imperative, such shifts can be an effective way to advance neoliberalization's agenda of outsourcing government decisions to quasi-governmental and private sector parties. If the state can be successfully tarred as the enemy of democracy, then outsourcing can be sold as democratization. Of course neither outsourcing nor devolution of authority is in itself necessarily a move toward greater democracy (Purcell 2006). It may be true that in the Keynesian era policymaking tended to be bureaucratized and undemocratic. But when that authority is ceded by the national state to local authorities or to non-state entities, it can be mobilized democratically or not. Devolution and outsourcing may or may not be democratization, but neoliberals sell it as such.

Given neoliberalism's penchant for adopting strategies of democratic legitimation, and given their strong insistence on individual liberty, it should come as no surprise that neoliberals regularly couple their agenda with that of liberal democracy. The connection with liberalism is made clear by the 1947 founding statement of the Mont Pelerin Society, the seminal group of neoliberal thinkers. It lays out clearly a distinctly liberal ethic when it worries that

> the position of the individual and the voluntary group are progressively undermined by extensions of arbitrary power [which, for the Society, means the Keynesian state]. Even that most precious possession of Western Man, freedom of thought and expression, is threatened by the spread of creeds which, claiming the privilege of tolerance when in the position of a minority, seek only to establish

a position of power in which they can suppress and obliterate all views but their own.

(Mont Pelerin Society 1947)

The statement then sutures the survival of such liberal values to the values of neoliberalism. It expresses alarm about "a decline of belief in private property and the competitive market; for without the diffused power and initiative associated with these institutions it is difficult to imagine a society in which freedom may be effectively preserved." Without unfettered capitalism, in other words, there can be no freedom. The Cold War era helped solidify that ideological claim: capitalism and liberal democracy, the belief was, stood against communism and totalitarian tyranny. Capitalism and democracy were presented as an essential pair, linked by freedom, that naturally allied against all forms of tyranny. The present era is strikingly similar: the advance of capitalist social relations through neoliberal globalization is ideologically twinned with the emergence of nascent forms of liberal democracy. Just as Margaret Thatcher gushed that "there is no alternative" to neoliberal reforms, Francis Fukuyama declared "the end of history" and the final triumph of liberal democracy and capitalism. In announcing the "unabashed victory of economic and political liberalism," Fukuyama, ideological heir to the Mont Pelerin neoliberals, was mobilizing freedom as the connector: free elections are the natural pair of free markets (Fukuyama 1989, p. 3). Free citizens and free consumers are two sides of the same coin. By extension, then, socialism and communism are fundamentally incompatible with democracy (Fukuyama 1992).

Fukuyama penned his original argument as a state-department employee under the first Bush administration. The younger-Bush administration, under the influence of Paul Wolfowitz among others, has taken the argument much farther, into a messianic vision for transforming the world. In its National Security Strategy of the United States of America (Office of the President 2002, p. iv), the administration argued that "the great struggles of the twentieth century between liberty and totalitarianism ended with a decisive victory for the forces of freedom—and a single sustainable model for national success: freedom, democracy, and free enterprise." While the totalitarian nature of that statement ("a single sustainable model") is certainly ironic, the continuity with Fukuyama is clear: liberal democracy and capitalist free enterprise are inextricably bound up together to create the only imaginable way forward. Each cannot stand without the other, a claim that recalls one of the founding fathers of neoliberalism, Milton Friedman, who argued (1962, p. 8) "economic freedom is … an indispensable means toward the achievement of political freedom" (see also Hartz 1955). Bush's National Security Strategy (p. 27) even goes so far as to reduce the capitalism/democracy pair to a single entity when it celebrates "the basic values of free-market democracy." Democracy without capitalism, they imply, is unthinkable; the only option outside that pair is the tyranny of "anti-free market authoritarianism" (Office of the President 2006, p. 3).[8] But even though the capitalism–democracy connection is seen as natural and necessary, the strong neoconservative influence in the administration means that the Bush administration does not plan merely to look on as the world becomes more capitalist–democratic. Rather it "will actively work to bring the hope of democracy, development, free markets, and free trade to every corner of the world" (Office of the President 2002, p. v). The administration's concept of democracy is distinctly a liberal one, stressing the importance of free elections, individual rights, the "rule of law," and limited government. And its particularly Lockean (1988) version of the liberal heritage imagines the limits to government in a very neoliberal way; among the institutions in civil society that must be protected from government are "private property, independent business, and a market economy" (Office of the President 2006, p. 4). The National Security Strategy of the United States could easily have been written in Mont Pelerin.

Despite the copious historical evidence to the contrary, which I mention above, neoliberals have been largely successful in making the case that democracy and neoliberalism are necessary for each other. Any project to oppose neoliberalization through democratization must confront that fact. The neoliberal/liberal-democracy pair is currently the hegemonic model for organizing political economies. It really is seen by many as the only sustainable model for political economies. But even if it is currently dominant, it is not invincible. History has not ended. Neoliberalization is an incomplete process that struggles with internal contradictions, manifests its agenda unevenly, and produces unintended consequences (Larner 2000, p. 21). Alternative urban futures are altogether possible, but unless we can reclaim democracy as a path of resistance, neoliberalization may very well extinguish their light.

CHAPTER 2

THE MANY FACES OF DEMOCRACY

In January of 2005, all over Iraq, and in Europe and the United States as well, the index fingers of Iraqi citizens were stained with purple ink. The ink was earned by voting in the Iraqi elections, the first real elections in over 50 years. Much was made of the fingers in the U.S., they were considered a powerful symbol of democracy in Iraq, and, therefore, of the hope for stability and peace. When the President gave his State of the Union speech three days later, many Republican members of Congress waived stained purple fingers in celebration. While some commentators recognized that the election was only one small component of a robust democracy, many others conflated the purple fingers with the arrival of democracy. "All day yesterday," the *Washington Post* reported,

> voters held up their purple fingers in triumph. It was a new victory sign, maybe someday a peace sign, they hoped. It befitted the low-tech, hands-on feel of this election—democracy at its most basic and emotionally powerful. Democracy had marked them, touched them physically, and they hoped it would last forever.
>
> (Montgomery 2005, p. C1)

In fact, their fingers were purple because they had voted in an election. But in the reporter's imagination "*democracy* had marked them, touched them physically." As one critic put it, the American foreign-policy project of "democratization [had] become confused with elections" (Hoffmann 2006, p. 60). The Iraqi elections produced a widespread and giddy sense that democracy had "won." In the future we will look back on the event as "the arrival of democracy" in Iraq (Easterbrook 2005). "The terrorist and insurgents are violently opposed to democracy," the president argued, and so they worked to undermine the elections (Bush 2005b). Despite the naysayers, he claimed, Iraq is making progress toward democracy, "from tyranny, to liberation, to national elections, to the writing of a constitution, in the space of two-and-a-half years" (Bush 2005a).

More generally, in the West elections in faraway places are regularly the cause for pronouncements about the "triumph" of democracy. In Palestine in 2006, 30 Hamas militiamen turned in their guns to vote, prompting the *Daily Telegraph* to proclaim the "triumph of democracy over guns" (Butcher 2006). The 2006 election of Michelle Bachelet as Chile's president was hailed by U.S. Secretary of State Condoleezza Rice as "the triumph of democracy over a troubled history" (Associated Press 2006). When Vicente Fox was elected president in 2000, the BBC reported that President Clinton praised the election as a triumph for democracy. And in 1998 politicians in Spain celebrated elections in the not-so-faraway Basque region as a "victory for democracy" (BBC 1998). A very different view was expressed by Joseph Stiglitz, who declared the triumph of democracy when the WTO negotiations collapsed at Cancun because poorer countries refused to acquiesce to the free-trade demands of the richer ones (Stiglitz 2004). Despite Stiglitz' dissention, however, the prevailing sense of democracy is a very thin one: democracy's "rule of the people" happens when individuals express their preferences by voting in elections to select governing representatives. That thin, liberal view of democracy is currently dominant. Moreover, it is commonly paired with a neoliberal economic ideal. Free elections (democracy) must go hand-in-hand with free markets (capitalism). Each will fail without the support of the other. In what follows, I want to help reclaim democracy from these dominant assumptions. My aim is to show there is quite a lot more to democracy than this impoverished thin-liberal view. Moreover, I seek to thoroughly dismantle the idea that democracy and neoliberalism are necessary to each other. On the contrary, I argue that democratization is a very promising way to *resist* neoliberalization and pursue more hopeful urban futures.

An Essentialist Alternative

One obvious alternative to the neoliberal argument that democracy and capitalism are a natural pair is to argue the opposite, that democracy and capitalism are by definition antithetical. A tradition in both critical political economy (e.g. Goonewardena 2003; see also Luxemburg 1970) and neoclassical economics (Bernholz 2000) argues just that. "Democratic innovation," John Dryzek (1996, p. 34) contends, must take place "in the face of the structural and ideological constraints" that capitalism, and especially neoliberalism, present.

Bowles and Gintis (1986, p. 3) begin their classic book with a clear opposition between democracy and capitalism.

> This work is animated by a commitment to the progressive extension of people's capacity to govern their personal lives and social histories. Making good this commitment, we will argue, requires establishing a democratic social order and eliminating the central institutions of the capitalist economy. So stark an opposition between "capitalism" and "democracy," terms widely held jointly to characterize our society, may appear unwarranted. But we will maintain that no capitalist society today may reasonably be called democratic in the straightforward sense of securing personal liberty and rendering the exercise of power socially accountable.

For most writers in this tradition, the antithetical relation relies on a conception of democracy that is essentialist. They speak of democracy as though it possesses an authentic root meaning. There is a "real" democracy, and there are imposters. Most intimate that root meaning without examining it, assuming it is self-evident common knowledge, as with Bowles and Gintis' democracy "in the straightforward sense." Douglas Lummis (1997) does better in that he acknowledges that there are many ways to understand democracy. But he argues that most of those ways misunderstand the true nature of democracy. He rejects, for example, the Cold War idea that democracy is "not-communism," the free-market democracy of neoliberals, and the notion that democracy is equivalent to elections. Those are not democracy, he claims, but rather "twisted and hypocritical uses" of the word (1997, p. 15). True democracy, he argues, is *direct* democracy: the rule of all by all. It is a radical idea that involves great transformations in society. For Lummis, capitalism is entirely incompatible with democracy. Hardt and Negri (2004) echo Lummis' notion of an original, unspoiled democracy that we must rediscover. In the eighteenth century, they say,

> the concept of democracy was not corrupted as it is now. The eighteenth century revolutionaries did not call democracy either the rule of a vanguard party or the rule of elected officials who are occasionally and in limited ways accountable to the multitude. They knew that democracy is a radical, absolute proposition that requires the rule of everyone by everyone.
>
> (2004, p. 307)

"We want nothing less than democracy," they write, "real democracy" (2004, p. 306).

They agree with Lummis that democracy and capitalism are fundamentally opposed. That assumption means that for such thinkers the way forward is clear:

democratization is *by definition* resistance to capitalism and neoliberalization. In this tradition one need not argue about the proper content of democratization. Rather we must rediscover "real" democracy and make it happen.

I argue that this counter-essentialism, while useful for undermining the purported synergy between neoliberalism and democracy, is not the best tack to take. Rather we should reject essentialist approaches to democracy. An essentialist understanding ignores or wishes away the fact of a very real political struggle over the meaning of democracy. It denies that democracy is, in fact, a malleable concept mobilized to pursue a political agenda, and treats it instead as an already-formed entity, one we must rediscover rather than reinvent for ourselves. Even if there were a "true" democracy, it is hard to avoid the philosophical skeptic's epistemological point that we have no way of *knowing for sure* which democracy is the true democracy (Landesman and Meeks 2002). From that point of view, any assertion of ontological essence, whether from the right or the left, is in fairly bad faith. Because we don't know what the true form is, or if there even is a true form, we are always unavoidably *making an argument* for our particular form of democracy over and above other forms. In making that argument, it is more to the point to say our idea of democracy is more just, more egalitarian, more civilized—or more democratic—than other forms, than it is to say our democracy is "real" and others are false. So, to be clear, by rejecting essentialism I do *not* mean to say that all forms of democracy are equally desirable. Not at all. While the "free-market democracy" of the neoliberals is an accepted (and indeed hegemonic) form of democracy, I argue that it is nevertheless entirely impoverished. There are far better options available. But if we are sidetracked into a debate about which democracy is the "real" democracy, we will have missed the opportunity to say, loud and clear, why free-market democracy is impoverished, and why an alternative democracy is more desirable. To do this, we absolutely must topple the current assumption that democracy and neoliberalism are fundamentally symbiotic, but we should do so by denying that such necessary symbioses exist at all.

Another reason to reject the essentialist approach is strategic. The antithesis argument isn't likely to go anywhere because it reads as quite weak in the context of observed facts. If there really *is* a "true" democracy, common sense would tell us it must be the one that has emerged now, as dominant at "the end of history," not an ancient one that got lost long ago. Given the current hegemony of neoliberalism and liberal democracy, positing a true democracy seems to cede the debate to the hegemonic form, giving it a powerful tool for its own reinforcement. However, there isn't any reason to wade into such waters. The anti-essentialist position is a more effective way to undermine the dominant notion of democracy. If there is no essence to democracy, then the neoliberal claim that there is a necessary connection between democracy and capitalism falls apart. And the anti-essentialist claim is bolstered by observed facts. While "free-market democracy" is dominant, it has not been dominant long, and Fukuyama's claim, while celebrated, ignores both the existence of many forms in the past and the currently proliferating democratic alternatives to liberal democracy. Democracy is, and has always been, a contested concept. Any apparent "end of history" is always a partial, and temporary, hegemony of one interpretation over others (Mouffe 2005). To unseat that hegemony, and advance an alternative, we need only make the case that alternative forms are preferable to actually existing liberal forms. Certainly that case is also an uphill climb, given the dominance of the liberal form. But it is less uphill that an essentialist challenge. And the considerable thinness of actually existing liberal democracy make it entirely vulnerable to arguments that a better democracy is possible.

Lastly, we should avoid the essentialist approach because it is likely to generate destructive dogmatic splits within the movement to resist neoliberalism. When we see democracy as a thing with a true essence, we are far more likely to splinter over ontology and descend into unproductive sniping, or worse, purging. Seeing democracy instead as an open and evolving frame in which to move forward would strongly mitigate that danger. It would not mean the end of internal debates, nor should it, but the more open approach does help prevent wars of existence within movements for alternative democracy.[1] So my approach is to reject an essentialist understanding of democracy in favor of an explicit critical examination of the various democratic forms. My goal is to get specific about the content of the democracy I advocate, to demonstrate why it is different from traditional liberal democracy, and to develop why it offers a better alternative to the neoliberal urban future. If we accept that democracy has no natural pair, then we can free ourselves from the tyranny of the neoliberal maxim that democracy and capitalism need each other, that without free markets there can be no free people, and vice versa. That freedom opens the ideological terrain to a set of democratic attitudes that oppose neoliberalization. Of course, the challenge becomes to articulate and mobilize those attitudes. In my approach, democratization becomes a distinct political project that must engage with and outmaneuver other political projects (such as that of the Bush administration). Because it rejects an essentialist approach, the project must be explicit about how its democratic attitudes oppose neoliberalization. Such specificity is particularly important given neoliberalism's tendency to absorb democratic movements. Unless a democratic movement clearly stands against neoliberalization, it can quickly be turned to neoliberal ends.

In addition to destabilizing neoliberalism's misleading assumptions about democracy, we must also reclaim the field of democratic discourse. Recall Larner's (2000) argument that neoliberalism is as much an ideology and hegemonic discourse as it is a set of policy options. The same can be said of democracy. It is not merely a way to structure decision-making. It is also an ideology and a discursive frame. The ideological power of democracy is extremely strong; most people accept it without question as good. But that power is currently being wielded more by neoliberals than by the resistance. Democracy's discursive and ideological power can be turned to resist rather than reinforce neoliberalism, but only if we can develop and promote a set of new and more radically popular democratic attitudes. This chapter argues that we should not invent those democratic attitudes from scratch but should forge them out of the wisdom of existing democratic traditions. The democratic attitudes I pose specify clear democratic values, but they are also malleable enough to be useful in a variety of different contexts: different cities, different policy arenas, and different neoliberal challenges. Neoliberalization is constantly recalibrating to resolve contractions and overcome challenges; the democratic resistance must be strategically flexible as well. For example, the democratic attitudes that follow take no absolute position on the role of the state. They do not rule out engagement with and even institutionalization in the state apparatus (Fung and Wright 2003). As we have seen, the state is fully imbricated in the project of neoliberalization. Moreover, the state itself involves problematic structures of domination and coercion (Day 2005; Graeber 2004). So we must always be cautious about engaging the state. At the same time, we should be open to and learn from the creative ways that democratic movements are engaging the state strategically to advance democratization and codify their gains. There are opportunities there, as well as dangers.[2]

So if we are to take advantage of the rich heritage of democratic thought in forging a new set of democratic attitudes, we must begin by critically examining that heritage. This chapter begins by reviewing contemporary democratic theory in order to produce a critical survey of the potential of existing democratic ideas. The way I organize that survey is by using the shorthand of "democratic traditions." Those traditions—liberal, deliberative, participatory, etc.—are compartmentalized here as a way to impose some cognitive order on the diversity of thought in democratic theory. The reality of the debates is of course more complex than the following account will be able to show. Each tradition features much internal debate, and the lines between, say, civic republicanism, communitarianism, deliberative democracy, and participatory democracy are sometimes clear, and sometimes shifting, and sometimes extremely blurred. Debates erupt over such labels; authors are placed in one camp and then vigorously resist the designation. My characterization of the camps might not match exactly that of other observers. The labels I have provided are merely what I consider a useful rubric for referring to the bulk of authors who take one or the other side of a debate. To help readers sort through the multiplicity of approaches, Table 2.1 summarizes some of the main features of each tradition. Once I have completed my examination of the democratic traditions, in Chapter 3 I draw on that examination to articulate the democratic attitudes I hope can help resist neoliberalization and democratize urban political economies.

Table 2.1 Summary of democratic traditions

Democratic Tradition	Essence of Democracy	Relationship to neoliberalism	Role of the state	Public/Private sphere	Equality	Stance on the common good
Liberal	Popular sovereignty that must be tamed by the need to protect the freedom of individuals	Strongly compatible	Neutral guarantor of individual rights	Strong division between public (state) and private (non-state) spheres in order to protect individual liberty	Formal political equality in public (state) sphere only	Individual interests are paramount; collective will is achieved by aggregating individual wills
Deliberative	Citizens deliberate toward shared understandings	Potentially strongly compatible	Neutral guarantor of proper deliberative procedures	Desire to extend democratic deliberation beyond narrow liberal limits and beyond state	Equality sought when citizens are deliberating	Common good is the animating logic of decision-making
Participatory	Citizens develop their civic virtue through political participation	Historically oppositional	Neutral but charged with encouraging political participation among citizens	Strong commitment to progressively extend democratic participation beyond state to all spheres of life	All participants are equal; those with more experience can help educate those with less	Common good is the animating logic of decision-making
Revolutionary	The people reclaim the power that has been usurped by particular interests	Entirely antithetical	Entirely transformed or eliminated as people assume direct power	Strong commitment to a total democracy that eliminates public/private distinction	Everyone is equally a member of 'the people'	Common good is the animating logic of decision-making
Radical Pluralism	Irreducible political difference engaged through agonistic struggle	Makes possible and explicitly encourages counter-hegemonic resistance to neoliberalism	Will likely (but not necessarily) play a key role in any hegemonic stabilization of power	Commitment to radically extend democratic relations to more spheres, though not to 'total democracy'	Commitment to radical equalization, though not to total equality	Particular interests should not be eliminated; they should be mobilized to pursue agonistic struggle

LIBERAL DEMOCRACY

Any investigation of liberal democracy always points up the existence of a vast diversity of liberal-democratic ideas (Smith 1987). "The liberal tradition," Kwame Anthony Appiah (2004, p. ix) writes, "is not so much a body of doctrine as a set of debates." "Still," he continues, "it is widely agreed that there *is* such a tradition." It is possible to sketch some basic liberal-democratic values, even if the shape and extent of those values are subject to much debate.[3] The best place to begin is to recognize that liberal democracy is essentially the fusing of two traditions that sit together only uneasily. The liberal aspect stresses individual freedom from tyranny; the democratic aspect stresses the importance of democratic control, of a body of citizens choosing their collective fate (Sartori 1987). Democracy urges us to maximize the power of the people, but liberalism requires us to curb that power. Except in the unlikely case of unanimity, all democratic decisions will inevitably impose themselves to some degree on a minority of the populace, and thus run afoul of the liberal insistence on freedom. As Mill (1998[1859]) put it, liberal democracy creates the need for the people "to limit their power over themselves." Mill's words suggest that in general liberal democracy is an attempt to use liberal values to tame and contain democracy. Those that critique liberal democracy are usually trying to undo liberalism's constraints and seek a more fully democratic alternative.

Liberal democrats stress the individual as the basic political unit. They are centrally concerned to protect that individual from tyranny. In a democracy, liberals worry, the minority is vulnerable to the "tyranny of the majority" exercised through state power (Mill 1976; de Tocqueville 2004). Liberal democracy is therefore in part designed to protect the individual from intrusion and persecution by the democratic state. In order to ensure that protection, individuals are invested with robust personal rights. They have liberties like speech and conscience, and they have rights like equal treatment before the law, a fair trial, and protection from unreasonable searches and seizures. Those individual rights are imagined to be inalienable. That is, once they are recognized, rights inhere in the individual: they cannot be sold or traded away. That does not mean, however, that they are pregiven or unchanging. On the contrary, rights only come into existence when they are claimed by conscious struggle. The body of liberal rights has therefore always been highly malleable because its proper extent is continually contested. The political history of liberal democracy has mostly involved the progressive expansion and extension of rights. Rights have been extended from white male property owners to other groups in society. That extension has been remarkable but is certainly not complete, as evidenced by the current struggle to secure the right of gay couples to marry. New rights, such as the right to organize or be educated or vote have been fought for and won (though never permanently). And still other new rights are currently being struggled over, such as or the right of all people to affordable health care.

One important fulcrum around which struggles over rights take place concerns the distinction between negative and positive liberty (Berlin 1969; de Tocqueville 2004). Negative liberties have to do with freedom from interference that prevents individuals from doing what they want. Personal rights that protect the individual from coercion, for example, are negative liberties (e.g. a right to worship or speak freely). Positive liberties, on the other hand, concern the individual's freedom to reach her fullest potential. They ensure provision of people's basic needs so that they are free to flourish to the best of their abilities. That kind of liberty is much more robust and requires much more of the political community to ensure it. Provisions like a minimum wage, free schooling, adequate nutrition, or affordable health care enable individuals to flourish. Much of the expansion of rights in recent times has been an attempt to go beyond rights for negative liberty and forge rights that ensure positive liberty. While the notion of positive liberty is embraced to an extent by some liberals (e.g. Macpherson 1973), positive liberties and the emphasis on personal growth and development are really more at home in the participatory democracy tradition (see below). The home ground of liberalism (and liberal democracy) has always been much more a negative conception of liberty, with its concern to protect individuals from state tyranny.

Individuals and their freedom are further protected in liberal democracy by a strong division between public and private spheres. Liberal democracies vigorously defend the idea that some issues are matters of public concern, and others are not. Within the private sphere, individuals are protected from state intrusion (negative liberty) and free to act according to their conscience. The liberal desire to maximize that protection means they believe the private sphere should be very large, even if they debate just how large it should be. A large private sphere is also important for the liberal values of tolerance and pluralism. Within the private sphere, liberals believe, a whole wealth of beliefs and practices should be tolerated, and a plural society should be nurtured. A unified vision of the good life (as civic republicans such as Sandel (1996) might advocate) is antithetical to liberal democracy; the state should remain neutral on such questions (Appiah 2004).

In a less progressive way, a large private sphere not only protects diverse beliefs, it can also be a very effective way to insulate the economy from democratization. Most liberal democrats argue that the economy is properly part of the private sphere and therefore not subject to democratic control. According to that logic, it is acceptable for capitalist corporations to be governed by hierarchical decision-making and maintain authoritarian relations between capital and labor. In this context, property rights are an especially relevant liberal concern. John Locke, in particular, was keen to protect the rights of property owners from the interventions of the state (Locke 1988). The case of struggles over urban land is instructive. In most cities the state has significant power to shape what is done with a parcel of land (zoning, building codes, environmental regulations, etc.). In a liberal democracy, a strong form of property rights is meant to protect the owner from too much state intervention. He or she has rights that limit how much influence the state can have over land use decisions. Frequently in controversial development cases, community groups want something very different from what a developer has proposed. When they appeal to their elected officials, however, they are often told government can do nothing because the proposed development meets all the codes, and the development company has the right to pursue whatever use they want. So liberal democracy's public–private separation helps limit the state's ability to control capital. That strong laissez-faire tendency in liberal democracy is of course strongly valued by neoliberals.

In liberal democracy, the state factors into the distinction between public and private in a very particular way. For most liberal democrats, the state is the arena in which public affairs are largely contained. In other democratic theories, democracy is seen quite comprehensively as a far-reaching way of life that should define all human relations. John Dewey (2004), to take a famous example, saw democracy as much more than a form of government, but as a "mode of associated living." Unlike Dewey, liberal democracy is much more willing to limit democracy to a form of government, and to equate the state and the public sphere. As Frank Cunningham (2001, p. 42) argues, "the majority of champions of liberal democracy as well as most of its critics think of it as mainly or even exclusively a matter of government in modern nation states." Its politics, Cunningham continues (2001, p. 43), have "to do with relations between a state … and people subject to its authority. Moreover, to be properly democratic, this relation between state and citizen must be in accord with formal democratic procedures." Elections for government representatives take on great importance in this light, since they constitute a central element of the relation between state and citizen. For liberal democrats, democracy is primarily seen as a form of government, and since elections choose that government, elections are the very stuff of democratic politics. Liberals give virtually all their attention to questions of free and fair electoral procedures and transparent and accountable representation because they are for them the basic stuff of democracy. The conflation of elections and democracy that we met earlier is therefore the manifestation of a particularly liberal interpretation of democracy.

Another reason elections play such a prominent role in liberal democracy has to do with how it tends to conceive of political preferences. Because of the focus on the individual, each person is imagined to have a set of interests that manifest themselves as preferences when the person chooses among political options. Thus a parent with children in the public schools might benefit from government investing more in education, and so s/he would prefer candidates who seemed likely to ensure such spending. While liberals acknowledge that all people are influenced by interactions with others in their society (Kymlicka 1990), they nevertheless stress that individuals should have autonomy to decide for themselves what their interests are and to form concomitant preferences. Liberals who follow Mill want individuals to rely on their own judgment because they see forming preferences as part of the process of personal growth (Appiah 2004). Liberals who value the autonomy of individuals as an end itself see forming preferences as a private matter that should not be subject to coercion or even persuasion by others. Thus interests and preferences are imagined to be private and quite atomized. Even if they are unavoidably influenced by interactions with others, political preferences should as much as possible belong to each individual alone. We can see that liberal sensibility manifest in cultural norms, such as the idea that it is rude to ask someone who they voted for. It is apparent in legal restrictions such as the prohibition of partisan campaigning within a specified distance of a polling place. And it is evident in the micro-geography of polling stations, where people vote in individual booths separated by partitions or even fully enveloping curtains. The will of the people as a whole, according to this logic, is determined by the aggregated preferences of all individual citizens in the polity (Levine 1981).[4] Voting thus complements closely the liberal understanding of politics. It allows each citizen to vote for their individual preference, the votes are counted, and the aggregate preference of "the people" is revealed. Liberal democracy thus offers a quite atomized vision of politics, relative to other traditions.

There is a strong sense in liberal democracy that individual preferences should properly be self-interested. That is, the individual should prefer political outcomes that are in her own best interest, not ones that are in the interest of the polity as a whole. One reason for such self-interest is the self-development ethic we saw above: for liberalism the highest calling of humans is, as Mill (1998[1859]) saw it, to bring "themselves nearer to the best thing they can be." As the primary good, self-development takes precedence over other concerns, such as the greater good. Another argument for the primacy of self-interest is more willing to embrace wider social benefits. Paralleling the neoliberal argument about free markets, some liberal democrats who adhere to social choice precepts argue that in a democracy it is right for individuals to prefer their self-interest, because those preferences will be aggregated by the democratic process and thereby ensure the general interest of the polity as a whole (Riley 1988; Peterson 1981; Downs 1957; Tullock 1970).

It is important to acknowledge that I have overdrawn these two points about atomization and self-interest just a bit in order to better contrast them later with other democratic forms. As with all other elements of this sketch there is debate with the liberal-democratic tradition about atomization and self-interest. The purest form of the aggregative model has its roots in the works of Arrow (1951), Schumpeter (1947), and Downs (1957) and their heirs in the social choice tradition. Most liberals acknowledge the importance of social relations and constraints in building an individual's identity and forming their preferences. Some insist more than others on individuals pursuing their self-interest. So my point here is to stress that *relative to other traditions*, the liberal-democratic view tends much more strongly toward atomized formation of individual preferences and their expression and aggregation through electoral processes.

One last element of liberal democracy concerns equality. All democratic traditions value equality highly. Democratic citizens are supposed to have equal access to and play equal roles in political decisions. But in many ways equality conflicts with liberty. Robert Nozick (1974) goes so far as to argue that liberalism necessitates anti-egalitarianism. One's freedom to succeed or fail in a society, to earn more or less consideration or wealth or respect, can't be squared easily with the need for equality. That conflict is especially severe in a capitalist economy, which systematically produces social inequalities. Redressing those inequalities means collectivist strategies that will intrude on the liberty of individuals (or firms) by doing things like collecting taxes to redistribute wealth, enforcing minimum wages, or requiring all children to go to school. In order to tame democracy's thirst for equality, liberal democrats tend to limit it. In order to have a healthy democracy, they argue, we need only create *formal political* equality: equal citizenship, equal political rights, equal representation, etc. (Berlin 1969). In the public sphere of formal state institutions, everyone enjoys formally equal rights, respect, responsibilities, and access. Some liberal democrats are inclined, with Mill, to argue that political equality requires some measure of social, educational, and economic equality as well (Dworkin 1983; Rawls 1993). But such positions begin to call into question one's commitment to liberalism, and thus one's label as a liberal democrat. For the most part liberal democrats place questions about more robust forms of social equality *outside* the public sphere, and they are very wary of public intervention in that sphere, even in the name of greater social equality. As we saw, however, claiming formal political equality in the face of manifest social inequality can often raise legitimacy problems for liberal democracies.

DELIBERATIVE DEMOCRACY

Deliberative democrats offer a critique of and alternative to many of the precepts of liberal democracy. Their principal intellectual inspiration is Jürgen Habermas' (1984) theory of communicative action. Joshua Cohen (1997) and John Dryzek (1990) are examples of adherents in political theory. In planning, the communicative ideal that Innes seeks is equated explicitly with Habermas' notion of communicative rationality (Innes 1996, 1998). Habermas's goal is to fashion a politics where outcomes are not determined by greater power, coercion, or force, but by what he calls "the forceless force of the better argument" (Habermas 1999, p. 450). It is politics through communication and discussion, not violence. The project puts him squarely in line with enlightenment ideals, especially those espoused by Immanuel Kant in his *Perpetual Peace* (2003). For Habermas' generation, the search for a way to eliminate violence from politics was not merely philosophical, but a matter of extreme practical urgency. It is in many ways a reaction to the horrors of the two World Wars. In discussing the modernist "dream of substituting reason for force," James Traub (2006, p. 7) argues that statesmen used to have

> little patience for such castles in the sky—until the nineteenth-century tangle of treaty obligations was buried, along with the youth of Europe, beneath the fields of Flanders. War had now become so abhorrent that the Kantian project seemed not naïve but indispensable … .

Habermas, born in Düsseldorf in 1929, also lived through Nazism—and so he had abundant motivation to discover a way to replace force with reason.

Habermas advocates "communicative rationality" through which people produce progressively more intersubjective understanding of the world through effective communication. If people can communicate in the right way, he believes, they can create a shared understanding that can allow for deep-seated and stable agreements, agreements that create a solid sense of shared purpose, rather than the antagonism that ends in violence. Bent Flyvbjerg (1998b, p. 188) suggests that for Habermas such ideal communication requires several things to be achieved: (1) no one who is affected by the decision is excluded from the discussion, (2) everyone in the conversation has an equal chance to speak, (3) each interlocutor is willing to empathize with and take seriously all others, (4) pre-existing power differences are prevented from distorting the conversation, and (5) participants must be pursuing not their own self-interest, but the common good of the group. The goal is for participants to use rational (that is, reason-giving) argument to move toward a shared understanding of the common good. Once they arrive at a shared common good, they can make the decisions that best serve that good. As Selya Benhabib (1994, p. 30) puts it, the goal is to create a polity in which "what is considered in the common interest of all results from processes of collective deliberation conducted rationally and fairly among free and equal individuals."

Habermas' approach relies heavily on rational thought and argument. But he understands rationality in a very particular way. He contrasts communicative rationality and communicative action with instrumental rationality and strategic action. In the latter two, participants pursue their self-interest, using rational argument not to reach a shared understanding of the common good, but as a tool to get what they want. He associates this rationality with industrial capitalism and its debilitating effect on social life. Worse still, such rationality could support the use of force or violence to achieve ends. Habermas stands against such instrumental rationality and self-interested politics. Habermas' communicative rationality urges participants to set aside their self-interest and instead work with others to intersubjectively develop a sense of the common good (Habermas 1990). Accordingly, Habermas is concerned more with process than with outcomes. If people reason together toward a shared understanding of the common good, and they do it according to the requirements of communicative rationality, then the content of that common good is less important than the fact that the decision is democratic and it is in the perceived best interests of all participants. In order to create such politics, Habermas believes, it is important to create a set of formal democratic procedures and institutionalize them so that communicative action can become the norm for political decisions. He is therefore very interested in the writing of constitutions as a way to establish *a priori* the norms of communicative rationality for a state apparatus. If the right procedures for democratic deliberation can be codified, he believes, we can create a polity in which rational argumentation and intersubjective understanding, not coercion or domination, is the basis for decision-making.

Deliberative democracy has many adherents both in political theory and in many disciplines beyond (e.g. Cohen 1997; Gutmann and Thompson 2004; Dryzek 2000; Parker 2003; Benhabib 1996; Bohman and Rehg 1997; Gastil 1993). However, its impact on urban governance is perhaps most keenly felt in the enthusiasm for it that urban planners have shown (e.g. Healey 1992; Innes 1995; Forester 1998; Susskind and Podziba 1999; McKearnan and Fairman 1999; Innes and Booher 1999). The particular manifestation of deliberative democracy in planning is usually called "communicative" or "collaborative" planning, and there is some truth to Innes' (1995) claim that this model is emerging as the central paradigm in planning theory. Both the more general tradition of deliberative democracy and its more specific manifestation of communicative planning are similar to liberal democracy in that they are marked by significant debates. Any sketch of their principles therefore necessarily must simplify complexity to a degree. Nevertheless, there is wide agreement on several key values, especially with respect to how deliberative approaches differ from the liberal standard.

The first principle is clear from Habermas' emphasis on communication. For deliberative democrats the very stuff of democracy is communication and argument among democratic citizens. They strongly oppose the liberal-democratic idea of an atomistic individual. They stand against the notion that individuals should form their preferences autonomously and then express those preferences in segregated polling stations. Instead, deliberative democrats believe that democratic debate should play a *constitutive* role in people's preference formation. An example of that vision is a "deliberative poll" in which preferences are not expressed or fully formed until participants have deliberated meaningfully about the issue at hand (Fishkin 1997). Citizens should have their own ideas, to be sure, but those ideas should be seen as provisional, as very much subject to change as they encounter the arguments of other citizens. That precept is the heir to Habermas' insistence on empathy among interlocutors: not only must everyone have the chance to speak in deliberations, but also all participants must commit to taking their peers seriously, to being genuinely open to the possibility that another's argument will change their mind. The liberal idea that individuals should rely primarily on their own judgment in formulating their preferences, for deliberative democrats, misses the communicative essence of democracy.

Moreover, similarly wrong is the liberal idea that each person's preferences should be expressed in isolation in a voting booth, and aggregated with other individual preferences to determine the will of the majority (Sunstein 1997). Such electoral decision-making also misses the point. Instead, deliberative democrats stress consensus-building and intersubjective agreement. They want citizens to reason together toward a shared definition of the problem and agree as a whole on the right solution. Each individual should not express his preferences privately but should be required to defend them to other citizens with reasons (Cohen 1997). For deliberative democrats, the liberal electoral model leaves each citizen unaccountable to her peers; she can vote how she wants and never has to justify her choice. In deliberative democracy, she must express her preference in public, give reasons to justify that preference, and be genuinely open to having that preference altered by the arguments of others (Gutmann and Thompson 2004). "A well-functioning system of democracy," Cass Sunstein (1997, p. 94) argues, "rests not on preferences but on reasons." Thus one important contribution of deliberative democrats is to profoundly unsettle the facile liberal conflation of elections and democracy.

Deliberative democrats also part ways with liberals on the issue of the public sphere. Deliberative democrats strongly desire that we extend communicative rationality to as many decision-making arenas as possible. That is, we should deliberate publicly about far more issues than liberals traditionally recognize as public. For example, while Gutmann and Thompson recognize that deliberation has limits, they nevertheless want the economy subject to much greater democratic deliberation. Citizens "need forums within which they can propose and debate issues concerning the basic economic structure of society, over which corporations exert a kind of control that is properly considered political, not only economic" (Gutmann and Thompson 2004, p. 34). John Gastil (1993) offers a detailed empirical examination of how deliberation can be applied in neighborhood councils, a co-op grocery store, and the most private of liberal arenas, the family. Others highlight the importance of democratizing all manner of institutions in civil society (Cohen and Rogers 1995).

As a result of this expanded conception of public decision-making, deliberative democrats are much more willing than liberals to see democratic politics as properly reaching beyond the state apparatus and its limited public sphere. Of course, they accept the importance of the state apparatus as a decision-making arena, and they want very much to see more communicative action inside the state (Habermas 2001). Flowing at least in part from Habermas' insistence on constitution writing, there remains a heavy focus on expanding deliberation within the state (Habermas 1994). Communicative planners, because planning is still largely carried out by the state, augment this tradition, although they also are beginning to focus more on the role non-state planners can play (Grengs 2002). But because of the expanded conception of what is properly public, some deliberative democrats like Gutmann, Thompson, Gastil, and Cohen also advocate deliberative-democratic decision-making outside the state apparatus.

On the question of state neutrality, deliberative democrats share more with liberal democrats. In facilitating communication among stakeholders, they believe the state should be neutral in the sense that it should not favor one group over another. For example, in acting as a facilitator the communicative planner should ensure that as participants build collective solutions, no one at the table is unfairly advantaged or disadvantaged, because that would distort communication and undermine the deliberative process. However, the state should *not* remain neutral on the question of what constitutes good communication and democratic decision-making. It should consciously reject the liberal ideal of leaving self-interested individuals alone to formulate their own preferences and instead bring them together to formulate collective solutions. The ground rules of deliberative communication mean the facilitator will actively intervene to ensure proper interaction, remind participants to defend their positions with reasons, encourage everyone to listen sincerely to others' arguments and be open to changing their preferences, and discourage people from giving self-interested rationales. For deliberative democrats, whenever deliberation occurs under state auspices, it will mean the state is actively encouraging deliberation and collective problem-solving over atomistic and aggregative decision-making, another significant departure from actually existing liberal democracy.

In terms of the content of that deliberation, deliberative democrats believe agreement is a goal to work toward, that disagreement should be progressively superceded by intersubjective agreement. It is too much to say that deliberative democrats demand unanimity, that they require nothing short of consensus. They are well aware that consensus is not always attainable, and they understand it is usually very hard work. But it is not wrong to say that they value agreement, they aim for consensus, and they want in general to "smooth out" the rough terrain of conflicting positions. Deliberative democrats consider discord an undesirable state that can and should be resolved into concord, and their applied work pursues that goal. They develop concrete ways to build consensus, resolve conflict, and facilitate agreement through dialogue. Perhaps the best example here is *The Consensus-Building Handbook: A Comprehensive Guide to Reaching Agreement* (Susskind *et al.* 1999). The 1,200-page book includes many of the leading figures in communicative planning. The first part offers expositions of various techniques planners and others can use to build consensus (stakeholder selection, facilitation, joint fact-finding, etc.). The second part describes 17 empirical cases in which consensus building was generally successful, usually because of the abilities of a skilled facilitator. A typical gambit in the cases is to celebrate the possibility of reaching agreement in the face of what Forester (1999) calls "deep value differences."

Forester's case in the book, for example, brought together Christian fundamentalists who thought being gay was a sin against God and gay activists who thought AIDS needed to be seen not as a moral issue but as a public health one. Their task was to formulate a planning document to shape state public health policy on AIDS. Through skillful mediation, the groups managed to agree on a policy statement. Forester does not mean to say that they resolved all their value differences; his point is that specific limited agreement between such opposing sides can be achieved if a skilled mediator can help the groups talk to each other in the most constructive way. Because of the importance of effective communication, the roles of mediator and facilitator take on monumental importance in this literature. Some advocates of communicative planning have gone so far as to announce the transformation of the role of the planner: from a rational, scientific *expert* who uses his expertise to produce the solution most in the public interest to a *facilitator* who helps a variety of interest groups deliberate together toward a shared solution (Innes 1995; also see Sandercock 1998). Deliberative democrats share with Habermas what can only be described as an abiding faith that communicative action can, in the long term, resolve virtually any conflict, and every participant can get what they want (Flyvbjerg 1998a). As Forester's celebrated facilitator puts it (1998, p. 217), the goal is to "build solutions together that accommodate, not compromise, the interests of everybody." Forester (1998, p. 220) stresses that we should interpret him as making "a vital practical distinction between jointly beneficial solutions that accommodate diverse interests and mere compromises that sacrifice those interests." Deliberative democrats seek resolutions that are "jointly beneficial" to everyone at the table.

In order to do that, the Habermasian communicative ideal requires much of participants in terms of how they interact. They must think in terms of the common interest rather that particular ones. Preferences should be expressed in terms of what is best for everyone, not what is best for the speaker or their group. Jane Mansbridge (1992; quoted in Sanders 1997, p. 360) writes that "when a society needs to discourage individual self-interest and encourage altruism, deliberation in public will often serve that end." Gutmann and Thompson (2004, p. 20) call this the capacity "to work together in a more cooperative 'first-person plural' spirit." One should not consider what is good for *me* but what is good for *us*. Recall from Chapter 1, Parker's (2003) purposefully reclaiming the Greek word "idiots" to describe self-interested people. "Any communication that cannot connect the particular to the general," writes John Dryzek (2000, p. 68), "should be excluded" from democratic deliberation. In deliberative discussions about water resources in the Florida Everglades, stakeholders encouraged each other to "leave your pet pigs at the door" when participants focused only on their particular interests (Fuller 2006). The goal in communicative rationality is to produce an intersubjective understanding of the common good. Arguments should therefore be about what each participant understands to be the common good.

A related directive is that participants should not use a position of greater power to influence decision-making. In Habermas' ideal speech act, participants must have equal power. That is, they must have an equal chance to offer rational argument that influences the group. Any pre-existing power differences, therefore, must be neutralized for the purposes of the deliberation. For example, deliberation cannot allow a person trained in debate to outmaneuver someone not so trained. It cannot allow a corporate representative to threaten the group with local plant closure. It cannot allow the search for intersubjective understanding to be distorted by power. Rather, what must carry the day is the better argument about what constitutes the common good. Here the facilitator again plays an important role. He or she must prevent participants from using power to influence the course of the discussion. Any such distortion would undermine the democratic character of communicative action. Moreover, participants must not be excluded, either overtly or in more subtle ways. Deliberations must involve all those affected by the decision. That means that people who do not traditionally participate in urban decision-making, because they are poor or not educated, must be included. Powerful interests should also participate, but they cannot carry more weight in the process than traditionally excluded groups. Moreover, all participants, regardless of their power, must have an equal chance to influence the discussion. All sorts of cultural values (and tendencies to devalue) must be set aside. Participants cannot value the arguments of the mayor more highly—just because he is the mayor—than they value those of a homeless person. They must judge the arguments of each impartially. Other tendencies toward cultural esteem, such as those based on race or gender or education, must similarly be suspended.

Because of the emphasis on reason, participants are encouraged to have a particular style. They should project a calm, reasoned, dispassionate manner. Many communication styles, like allusion, testimony, rhetoric, storytelling, hyperbole, bombast, and bullying are seen as distortions of communication because they appeal not to calm reason but to other emotional impulses. Some who classify themselves as deliberative democrats, such as Iris Marion Young (1996, p. 122), have warned that such stylistic values result in "the devaluation of some people's style of speech and the elevation of others." In other words, Young is worried calm reason favors educated white men over marginalized populations, and this systematically excludes those groups (Young 2000). Despite Young's sympathetic critique, however, deliberative democrats often see emotion and excitement as something to distrust and overcome, not as a legitimate way to give public reasons (Downs 1989). Forester's mediator, for example, stresses "the importance of selecting spokespeople … who will have a 'calm presence' in a group when emotions run high" (Forester 1999, p. 483). Most deliberative democrats are not rigid on this point. It is probably most accurate to say that the communicative ideal of non-distorted communication pushes deliberative democrats to prefer "cool reason-giving" (Gutmann and Thompson 2004, p. 50) because they perceive it to be less likely than other forms to produce misleading, manipulative, or coercive communication. Where they are less permissive is on the issue of "pet pigs": deliberative democrats very much insist that the "force of the better argument" lies in appeals to the common good, not to particular interests (Young 2000).

PARTICIPATORY DEMOCRACY

The participatory tradition is similar to deliberative democracy, especially in its critique of the liberal model. In many ways, it presents a more forceful critique. However, instead of deliberative democracy's focus on communication, participatory democrats, as the name implies, build their model around the virtue of political participation. The historical birth of the tradition is often thought to be the student movements of the 1960s and particularly the "Port Huron Statement" that came out of a convention of Students for a Democratic Society (SDS). "In a participatory democracy," the statement says, "politics has the function of bringing people out of isolation and into community" (Students for a Democratic Society 1962). It is quite clear then that participatory democrats from the beginning rejected the individual ethos of liberal democracy and wanted to foster instead each person's robust engagement in their community. Such engagement, SDS believed, is necessary for each person to find meaning in his or her personal and political life.

That rejection of individualism and emphasis on community reveals that the tradition has roots that reach back before 1962. The communitarian and civic republican traditions are important cognates to participatory democracy. The former is more contemporary. Its goal is to react directly to liberalism, shifting the focus of political life away from the individual and toward the community (Etzioni 1994; Bell 1993; Tam 1998). The latter has very long roots that stretch back at least to Rousseau.[5] It shares with communitarianism a stress on the importance of the common good over and above individual interests. Rousseau felt that individuals reach their full potential only when they choose to give themselves over to the pursuit of the common good (Rousseau 1987, pp. 141–227). Pursuing self-interest, for Rousseau, is instinctual, animalistic. Working together to govern themselves democratically allows people to reach a higher, more fully human existence. True freedom, for Rousseau, lies in *choosing* to subsume one's own interest to the common good. As Susan Dunn (2002, pp. 11–12) puts it,

> Rousseau was convinced that individuals, through their identification with the community, could attain something higher than private happiness: *meaningful lives*. No longer drowning in infernal rat-races, competing and striving to impress others, pursuing elusive, inauthentic goals that never lead to real satisfaction, people by choosing to live as parts of a whole and to share in the common happiness of all have a chance at real fulfillment and peace of mind.

That emphasis on the common has been taken up by modern civic republicans (e.g. Sandel 1996; Petit 1997; Skinner 1992). Both communitarianism and republicanism share a desire to build a strong, cohesive polity in which individuals can be actively engaged in pursuing their common interest.

Participatory democrats very much share that focus on the common good. However, Rousseau had a very strange way of conceiving of political participation that participatory democrats desire to go beyond (though see Pateman 1985). Rousseau believed all decisions should be made to advance the common good, but he felt discovering the common good would be a fairly straightforward procedure. Assuming citizens are well-informed and have the collective interest at heart, they merely meet to express their idea of that good. He anticipated there would be only small differences among the various understandings of the common good, and a clear picture would quickly emerge (Rousseau 1987, pp. 155–6). Lengthy deliberation, debate, and argument were not part of his imagined procedures. Participatory democrats, on the contrary, conceive of political life as *constituted* by long hours of robust political engagement and debate. They believe participation through debate is absolutely central to democracy and to the pursuit of the common good.

The centrality of participation traces its lineage to Aristotle. Whereas Rousseau thought that surrendering to the common good was the way to maximize one's human potential, Aristotle believed instead that engaging in politics was how humans could improve themselves. For Aristotle the purpose of a human being is to be happy. In order to be happy, s/he must be virtuous. In order to most fully develop one's virtue, one must engage in politics, which for Aristotle meant the collective pursuit of the common good (Aristotle 1962, 2004). There is a clear sense for Aristotle of a "good life," that some ways of living are better and more meaningful than others. "The main concern of politics," Aristotle (2004, 1099b30) writes, "is to engender a certain character in the citizens and to make them good and disposed to perform noble actions." Participatory democrats adopt Aristotle's developmental notion of participation. As citizens participate more and more in democratic life, they come closer and closer to realizing their best selves (Pateman 1985). One is tempted here to recall Mill's (1998[1859]) argument that each individual's highest calling is to bring "themselves nearer to the best thing they can be." For Mill "the best thing" is unique to each individual, and they should seek it by relying on their own judgment. For Aristotle and for participatory democrats, "the best thing" is to be a citizen engaged in political congress with one's peers. To realize their best self, citizens must develop their civic virtue, which is to say their capacity to discern and act in favor of the common interest. The goal of developing the civic virtue of citizens is a good in and of itself, and it is one key aim of participatory democracy.

Another key aim is tightly linked to the first. As citizens develop their civic virtue, they become wiser, more just, more capable decision-makers. They therefore become better at discerning and deciding in favor of the common good. The more a polity encourages and enables participation, the argument goes, the wiser the decisions of its citizens will be, to the benefit of the polity as a whole. So participation is not just for the development of each citizen, it is also good for the group. For participatory democrats the citizen apathy that we observe in most democracies is not the result of a lack of will or interest on the part of citizens, it is the result of a lack of opportunities for meaningful participation (Barber 2004; Verba and Nie 1972). It is the responsibility of the polity to provide those opportunities, for the sake of both the citizens and the polity itself.

Because participation is the heart of democracy, therefore, participatory democrats want to promote it wherever they can. They recoil at the liberal equation of voting and democracy. Voting is only the most basic way citizens participate; a healthy democracy requires vastly more than mere elections (Barber 2004; Mansbridge 1983). Most participatory democrats want more participation in all spheres, from the economy to the family to civil society (Gould 1988). Few see it as confined to the "public sphere" and the state apparatus the way liberals do. Participatory democrats want to go beyond the tentative moves of deliberative democrats to expand deliberation beyond the state (Pateman 1970; Bachrach 1967). Cunningham (2001, p. 127) argues that "for participatory democrats state and civil society are not distinct entities;

there is no line dividing a state that rules and citizens in civil society who are ruled." Recall Dewey's (2004, p. 83) notion of democracy as "a mode of associated living, of conjoint communicated experience." In liberal democracy, individuals are relatively autonomous agents that engage each other politically in very limited ways. In participatory democracy, individuals remain distinct, but their experience is "conjoint," tied together comprehensively and inextricably. Democracy is how they make their way together in the world. So participatory democrats tend to see the polity in Aristotle's sense: not a formal state apparatus, but a whole society, where politics encompasses all aspects of daily life. They therefore go much farther than deliberative democrats and argue for a weak (or even non-existent) distinction between the public and private spheres. Almost all human interaction bears on our "mode of associated living," and should therefore be a subject for democratic politics.

As a result of this emphasis on participation, participatory democrats tend to be hostile to most forms of representation (Levine 1981; Hirst 1994). In liberal theory representation is an acceptable practical accommodation to large publics. In fact, in some liberal theory representation can even be seen as a good because it allows specialized political representatives to see to public affairs while individual citizens tend to their private affairs. It thus helps maintain the separation between public and private and the autonomy of individuals. For deliberative democrats, the central question is about the right kind of communication. Representation is acceptable if the representatives are deliberating effectively. In fact, since representative government is so much the norm in modern democracies, as a practical matter it makes sense for deliberative democrats to work for more deliberation among representatives, which for them is better than no deliberation at all. For participatory democrats, however, participation is a good in itself. Representation subverts that good. A member of the U.S. Congress, for instance, represents a district of around 700,000 people. She and her staff carry out everyday political participation *in place of* the citizens in her district—attending meetings, engaging other representatives, learning about issues, arguing in favor of particular points of view, voting, etc. At their most active, her constituents might write her a letter urging her to support a particular position. Less actively, if they are eligible they might vote for or against her every two years. Probably a majority will not even do that.

Such a structure is designed to limit democratic participation to a small group of people. It reduces 700,000 citizens to largely passive spectators. It is the classic example of how apathy is bred by limited opportunities for participation. For participatory democrats, that system actively erodes the essence of a democracy. Citizens should be actively engaged in all aspects of political life, both in formal government and beyond.[6] The great challenge to such participation is practical. If people participated directly in every political decision they would have no time for other pursuits; everyone would need to be a professional politician.[7] In addition, in very large—e.g. national-scale—polities, it is not feasible to have everyone meet and engage each other. A national-scale town hall, even in a small country, isn't realistic. As a result, participatory democrats strongly favor very small political groupings (Gastil 1993). Small civic associations, co-op-sized businesses, and town-hall or even neighborhood-scale government are their ideals. Archon Fung's (2003) work on "empowered participatory governance," for example, features neighborhood-scale and school-district-scale participation in decisions about policing and education, respectively. Devolution of authority "down to the neighborhoods" is seen as a key element that can enable enough local control to create meaningful participatory structures (Fung 2004, pp. 31–68). Anything larger tends to necessitate a representative structure. Therefore, by extension, for most participatory democrats increasing the size of a polity necessarily undermines its democratic character (Cohen and Rogers 1995; Hirst 1994). A national-scale democracy is nearly impossible when participation is seen as the linchpin of democracy. Participatory democrats tend then to be localists, and they tend to pursue their vision in small-scale polities. The New England town meeting serves as their enduring archetype (Bryan 2003).

One relatively large small-scale on which participatory ideals have been pursued is the urban, and the experiment with "participatory budgeting" in Brazilian cities has been a much-discussed case study. Since about 1988 in Brazil, the Workers' Party (PT) has won electoral victories in municipal, state, and eventually national elections. One early victory was in the city of Porto Alegre in 1988. There they undertook an experiment whose goal was in part to empower the disenfranchised. They set up a system of participatory budgeting, whereby citizens were able to decide together in decentralized participatory forums how the city budget should allocate municipal revenue. The model was subsequently extended to other cities like Betim and Belo Horizonte. In Porto Alegre, each of 16 districts convenes a citizens' assembly. Those assemblies establish the budget priorities for their area and choose delegates who will represent the area at a citywide scale. Then the delegates deliberate toward a budget for the city as a whole. Here they are explicitly instructed to consider citywide priorities, as well as advocate for their own area. The budget that results is then submitted to the mayor, who can either sign or veto it (Baiocchi 2003). The process shares participatory democrats' idea of the two main benefits of participation: citizen development and wise public decisions. Because participants were often new to politics, the process involved a very strong developmental ideal in that political learning was consciously built in to the process. Expert consultants offered participants instruction on the specific details of the budgeting process, and citizens who had more experience participating were expected to be patient with and help educate newcomers. Part of that education involved a strong ethic, true to participatory ideals, that participants should argue for the common good rather than the good of their particular group (Baiocchi 2003, p. 56). There was also a strong sense that one outcome of the new model would be to improve the quality of governance in the city.

But the process also had another, more strategic motive. The assemblies actively recruited participants from traditionally excluded populations: the poor and less educated. For the PT the process was "primarily intended as a mean for previously excluded, ignored, and/or underserved nonelites to have access to, and to become more active participants in, democratic politics" (Nylen 2003, p. 62). Due in no small part to the participation of these groups, there has been an important redistribution of municipal spending toward services for poor and marginalized areas and populations. The particular outcomes of course vary from place to place and year to year. But the overall result has been budgets much more favorable to the disenfranchised, precisely as the PT had hoped. Those populations of course constitute the PT's political base, which is now politically mobilized and engaged. While the standard ideals of participatory democracy—citizen development and wise decision-making—were very much in play, there has been an important sense in Brazil that participation is also a means to achieve the end of greater social justice, at least at the urban scale. Indeed, in most polities, expanding participation means extending political influence to those who have been traditionally excluded from politics. If we accept Harvey's (2005, p. 119) argument that the project of neoliberalization is to restore "naked class power" to the hands of capitalists, then the agenda of expanding participation in decision-making will very often involve resisting the neoliberal project to some degree. The implied goal is to increase "popular," rather than elite, participation. That left-leaning (and potentially anti-neoliberal) content of participatory-democratic thought is fairly typical of the tradition. Even if participatory political philosophers don't necessarily make it explicit, it is worth being mindful of this more strategic agenda, for it can be, in practice, an important outcome of participatory democracy.

As the Brazil case makes clear, there is a great deal of overlap between deliberative and participatory democracy. While participatory democrats *tend* to imagine a wider role for democracy in society than do deliberative democrats, the latter also hope to extend deliberation beyond the state and the formal public sphere, even if to date they have tended to focus on the state. While they do differ in their respective emphases on deliberation and participation, it is possible to fuse the two by insisting that participation take the form of deliberation through communicative rationality. Participatory budgeting strove to achieve this kind of mixture, and it is not uncommon to see the two traditions work hand in hand. Both share a rejection of the atomistic and electoral tendencies of liberal democracy, both argue that politics should strive for consensus, and both maintain that the common good should guide decisions. That focus on the common good and the tendency to stress collective interests over individual (or small-group) ones has been the target of much liberal critique (Rawls 1993; see also McCarthy 1994; Benhabib 1996), as well as critiques from radical pluralists (whose ideas I discuss below). The most strident of liberal critiques point to liberals' greatest fear: the Rousseauian and civic-republican currents in participatory democracy, they worry, make it possible that participatory democracy is as likely to produce totalitarianism as it is democracy. As with many critics of Rousseau, they cite the history of the French Revolution as evidence of how Rousseau's ideals can go horribly wrong. However, one might argue instead that the concern about totalitarianism is even more germane to the next tradition, revolutionary democracy.

REVOLUTIONARY DEMOCRACY

The label "revolutionary democracy" is uncommon. Few in the tradition use the label. Lummis, for example, thinks of himself as a radical democrat, as do many in the tradition. However, I am concerned to avoid confusion with the next tradition, radical pluralism, and so I use the term revolutionary democracy to get at what I see as the main feature of this tradition: the argument that democratization requires not instituting elections or associations or a new way of talking, but rather it requires a full-blown transformation of society (see Marx 1977). "For Marx," as Michael Lowy (2003, p. viii) writes, "revolutionary democracy—the political equivalent of self-emancipation—was not an optional dimension but rather the intrinsic nature of socialism itself, as the free association of individuals who take into their hands the production of their common life." Democracy, for revolutionary democrats, entails nothing less than an economic, political, and cultural revolution.

To develop the arguments of the tradition, I will focus on two bodies of work, that of Douglas Lummis and Michael Hardt/Antonio Negri. We have met them already, in the discussion above about essentialism and democracy. But the roots of the revolutionary-democratic perspective run to various sources. Marx (1977) is in many ways the seminal figure. In Henri Lefebvre also, there are intimations that democracy must involve a thoroughgoing revolution in urban life (1968, 1986, 1991a, 2003). Rosa Luxemburg (1970, p. 391) advocated a socialism that incorporated "the most unlimited, the broadest democracy and public opinion." Many other more modern writers could also be included in this tradition (e.g. Lebowitz 2006; Lowy and Betto, no date; Wolin 1994; Howard 1990). Revolutionary democrats tend to conceive of democracy as having a true meaning, an original sense that, if realized, would transform society utterly. Against this background, current political systems that are labeled "democracy" are merely masquerading, to greater or lesser degrees. The spirit of this position was captured in 1871 by Walt Whitman. In his "Democratic Vistas" he wrote

> We have frequently printed the word Democracy, yet I cannot too often repeat that it is a word the real gist of which still sleeps, quite unawakened, notwithstanding the resonance and the many angry tempests out of which its syllables have come, from pen or tongue. It is a great word, whose history, I suppose, remains unwritten, because that history has yet to be enacted.
>
> (Whitman 2004, p. 248)

Democracy, for Whitman, has a "real gist" that is asleep. He locates that gist in the future, as something that will one day be enacted. Douglas Lummis (1997, p. 15) agrees that democracy "contains a promise yet to be fulfilled." However, for Lummis (1997, p. 26) the true gist of democracy lies in the past, in democracy's original meaning: "a democratic revolution is not a leap forward into the uncharted future; it is, as John Locke indicated, a going back, a return to the source." Lummis uses the term radical for himself, therefore, because he wants to return to the root meaning of democracy, its etymological source, and to cast aside the "misunderstandings and disfigurements" of the word. As we saw above, Hardt and Negri, though they are perhaps less strident, share Lummis' notion of a real democracy that has been corrupted.

For Hardt and Negri (2004, p. 317), democracy is "the rule of everyone by everyone." They go back to Polybius' distinction among three good forms of power: monarchy, aristocracy, and democracy (Polybius 1979; Hardt and Negri 2000, pp. 314–16). Monarchy is the rule of all by one, aristocracy is the rule of all by some, and democracy is the rule of all by many.[8] Hardt and Negri (2004, pp. 240–1) go on to explain that "the first great innovation of modern democracy" was to extend the ancients' notion of "the many" to include everyone.

> The ancient notion of democracy is a limited concept just as are monarchy and aristocracy: the many that rule is still only a portion of the entire social whole. Modern democracy, in contrast, has no limits and this is why Spinoza calls it "absolute." This move from the *many* to *everyone* is a small semantic shift, but one with extraordinarily radical consequences! With this universality come equally radical conceptions of equality and freedom. We can only all rule when we do so with equal powers, free to act and choose as each of us pleases.

For Hardt and Negri, then, modern democracy is the rule of all by *all*, rather than by many. Other forms of government that include monarchy or aristocracy, such as the U.S. model (President as monarch, Senate as aristocracy, and electorate as demos) is not properly democratic, but a hybrid form. In this light, representation, which Hardt and Negri call the second great innovation of modern democracy, allowing it to be extended to nation-state sized polities, is hard to square with democracy. It is important to recognize, they tell us, "that democracy and representation stand at odds with one another. When the power is transferred to a group of rulers, then we all no longer rule, we are separated from power and government" (2004, p. 244). Lummis agrees, arguing that "democracy is not a situation in which the people turn over their power to someone else in exchange for promises" (1997, p. 16).

While such a radical idea of direct and absolute democracy seems an impossible fantasy when compared with more pragmatic models like liberal democracy, it nevertheless contains a very important and compelling element of faith. That faith maintains that while revolutionary democracy may seem far-fetched, it is in fact more likely than it seems. True democracy, for revolutionary democrats, is not a pipe dream; it is rather the stable ground state of politics. Other forms are unstable deviations. In order to understand that belief we must understand a part of Marx's analysis of labor and value. Marx argued that labor is the source of all value, but the value that labor produces is appropriated by capital through a regime of social structures like property rights and wage relations. Capital therefore only exists as a parasite on the productive capacity of labor. That analysis is one reason Marx was so teleological about the advance of socialism. Capital can only rule through smoke and mirrors, because it does not itself produce wealth, and that simple fact of production will eventually bring down capital's house of cards (Hardt and Negri 2000, e.g. pp. 156–9). Marx's economic analysis of value is extended by both Hardt and Negri and Lummis into a more purely political one. They argue that the mass of people are in fact the source of all power. Any political system other than Spinoza's absolute democracy means that a few or a one must usurp the power that the people themselves generate. As with capitalism, that usurpation is ultimately an unstable arrangement because in the long run power will tend to return to its source, the people. While this faith may seem overly optimistic, it derives from a logical analysis of politics. The current regime "pretends to be the master of the world because it can destroy it. What a horrible illusion! In reality we are the masters of the world because our desire and labor regenerate it continuously" (Hardt and Negri 2000, p. 388).

However, warns Lummis (1997, pp. 25–6),

> the fact that the people are the source of all political power ... does not mean that in all regimes the people have the power, any more than the fact that workers are the source of all economic value means that in all economies the workers control the wealth. Every political regime is built by the taking of power from all the people and the giving of it to a few, every ideology is an explanation of why this power transfer is justified, and regimes are stable and powerful when the people accept those justifications.

That way of seeing the world produces a very clear political project: the people must reject the justifications of the regime and *reclaim* the power that is, in fact, theirs. In order to do so, "the people must form itself into a body by which power can in principle be held" (Lummis 1997, p. 21), and they must do this by making "the discovery that the real source of power is themselves" (p. 26). Once they become a body and act together to regain their power, democracy will be reborn, and the people will be "gathered in the public space, with neither the great paternal Leviathan nor the great maternal society standing over them, but only the empty sky—the people making the power of Leviathan their own again, free to speak, to choose, to act" (Lummis 1997, p. 27).

Hardt and Negri agree with Lummis that establishing democracy requires that power be returned to the source, and they agree that source of power is an active political agent. But they disagree in an important way with Lummis' way of conceiving that agent. Lummis' agent, "the people," is so unitary that Lummis refers to "it" in the third person singular rather than plural. Hardt and Negri, by contrast, explicitly reject "the people" as a way to characterize the agent. Instead, they use the term "multitude." "The people," they write (2004, p. xiv),

> has traditionally been a unitary conception. The population, of course, is characterized by all kinds of differences, but "the people" reduces that diversity to a unity and makes of the population a single identity: "the people" is one. The multitude, in contrast, is many. The multitude is composed of innumerable internal differences that can never be reduced to a unity or a single identity— different cultures, races, ethnicities, genders, and sexual orientations; different forms of labor; different ways of living; different views of the world; and different desires.

While their plural conception avoids the old problem on the left of stuffing significant differences into a one-size-fits-all movement, it leaves unsolved the basic political problem: the way for the dispossessed to regain their power is to come together, to act as though they are one even when they are many. Indeed, the fact of their multiplicity is the basis of divide-and-rule strategies that allow undemocratic regimes to appropriate their power. But the solution of forming a unitary subject has proven to be both an unpractical and an undesirable response. Much of Hardt and Negri's book is devoted to the task of constructing a conception of the multitude. It must be a political subject that can act in common—coming together around a common experience of being dominated by the current global regime[9]—but it must remain irreducibly and vitally differentiated. That common experience, the linchpin that allows the multitude to hold together, is not "discovered" as it is for Lummis; rather it is produced by the multitude through their social interaction and cooperation.

Hardt and Negri's vision is that the multitude will constitute itself through cooperative networks. Each local struggle against the regime can function "as a node that communicates with all other nodes without any hub or center of intelligence. Each struggle remains singular and tied to its local conditions but at the same time is immersed in the common web" of struggle against Empire (Hardt and Negri 2004, p. 217). That cooperative but decentered network of movements is a promising new model for social action and resistance, one that had its most prominent coming out at the demonstrations against the WTO in Seattle, where labor, environmentalists, and religious groups, among others, came together around a common—though not identical—opposition to the WTO's neoliberal project. Hardt and Negri's way of conceiving the multitude as a networked movement resonates in important ways with Laclau and Mouffe's (1985) concept of "chains of equivalence," which I explore in the next section.

Before I do, however, I must make one last important point about revolutionary democracy. Its vision is utterly incompatible with both capitalism and neoliberalism. While other traditions may call for the extension of their democratic values into the economic sphere (e.g. more deliberation in corporate governance), a revolutionary democracy, as it is conceived here, would necessitate the overthrow of the capitalist system. Some accommodation with capitalism, as liberal, deliberative, social, and even most participatory democrats allow, is inconceivable for revolutionary democracy. The only truly democratic political-economic system is one in which the producers of wealth control it and the source of all power holds it. Liberal democracy and neoliberal capitalism therefore have no place in the revolutionary-democratic imagination. That unequivocal rejection is important because it explicitly resists neoliberalization's ongoing project to absorb democratic practices. While neoliberalism has almost fully absorbed liberal democracy and may be in the process of effectively absorbing deliberative democracy, it is utterly unthinkable that revolutionary democracy could suffer the same fate. That kernel—explicit rejection of a neoliberal political economy—is among the elements of revolutionary democracy that I adopt in the democratic attitudes I develop below.

RADICAL PLURALISM

Radical pluralism is the newest of the traditions surveyed here. It is in many ways a critique of other forms of democratic thought. To date, radical pluralists have done less to chart what the radical pluralist alternative would look like. However, radical pluralism's critique opens important new ways forward for democratic practice, and they have offered more in the way of alternatives than is sometimes thought. This section aims to provide an account of the critique and then extrapolate from it what kind of democratic attitudes might result. The leading figure in radical pluralism has been Chantal Mouffe, but she is joined by her co-author Ernesto Laclau, as well as William Connolly and Claude Lefort. I also include here other authors who may or may not identify with the label (e.g. Nancy Fraser, Iris Marion Young, Bent Flyvbjerg, Leone Sandercock) because their have offered similarly trenchant critiques of the other forms of democracy. As such, together they constitute a body of critique that lies distinctly outside the other traditions. Moreover, each author shares a deep conviction that democracy requires a serious engagement with difference, conflict, and power.

To begin, it is important to distinguish between classic pluralism and radical pluralism. Classic pluralists, such as Robert Dahl (in his earlier work, e.g. 1967) and Nelson Polsby (1980) see themselves as liberal democrats, but rather than focus on the individual, they conceive of interest groups as the basic unit of politics. Interest groups form and mobilize their power to secure their particular interests against those of other self-interested groups. Classic pluralists therefore do not share the common-interest ethic of deliberative and participatory forms. Moreover, they accept that power and conflict will play a central role in political life. The focus on pluralism derives ultimately from the foundational concerns of James Madison. Societies are unavoidably marked by conflicts among interest groups, Madison believed, and so the question is how democracy can regulate that conflict so that it does not lead to the dissolution of the republic.

Radical pluralists very much take up this concern with managing conflict. However, they differ from classic pluralists in two important respects. First, classic pluralists saw power as relatively discrete, that is, groups can hold it independent of their relations with other groups. Each group works to amass the commodity of power, and then it applies it in the political arena (usually assumed to be the state, as befits their liberal heritage) to pursue its interests. Radical pluralists, on the contrary, see power and conflict not as discrete entities, but as constitutive of social relationships. Power only exists relationally; it inheres in the relations of dominance between two social groups. It is, therefore, an unavoidable and ubiquitous condition of social life, not a discrete and alienable asset that can be bracketed and laid aside. Second, classic pluralists see power as extremely diffused throughout society. The various interest groups are seen as more or less equal combatants for influence (Judge 1995; Harding 1995). Radical pluralists see power instead as chronically unequal. Power relations are typically characterized by hegemonies: a particular group or alliance of groups succeeds in establishing their political agenda as the dominant agenda in society as a whole. The notion of hegemony varies to a degree among radical pluralists, but it is broadly understood in a Gramscian sense to mean a temporary stabilization of power in which the interest of a particular group is accepted as the same thing as the interest of the whole society. The classic example of hegemony in that sense is when the capitalist classes successfully establish their particular interests as the common interest: "what's good for General Motors is good for America" as that company's president once declared. However, the structure of a given hegemony can involve a variety of social relations of domination, such as those associated with race, gender, age, language, etc. The same notion of hegemony can also be used to understand both the Keynesian compromise and the neoliberal alternative that replaced it. Any social relationship, for radical pluralists, will be necessarily constituted, at least in part, by the power relationships that flow from such hegemonies. While hegemonies are by definition never eternal and never total, nevertheless they can endure over time. Empirically some groups consistently dominate other groups, and their differential power defines their relations. For example, Mouffe's (1993, p. 10) concern for "the growing domination of relations of capitalist production" is a central concern for her democratic politics. Such concerns about hegemony did not trouble the minds of classic pluralists (although see Dahl 1985).

The second important distinction I should make is between radical pluralists and liberal democrats. Classical pluralists, as we just saw, tend to be liberals, and it is fair to claim that radical pluralists also have an important, though less significant, connection to liberal democracy. Mouffe herself wants to rescue the potential of liberal democratic theory from the current way it is being understood. She fears that "liberal democracy is increasingly identified with 'actually existing liberal democratic capitalism'" (1993, p. 6). The seed in liberalism that Mouffe wants to reclaim has to do with the way it handles pluralism and difference. Unlike deliberative, participatory, or revolutionary models, liberal democracy is much more inclined to accommodate and indeed champion diverse perspectives and political positions. The basis of Mouffe's argument is that difference, pluralism, and conflict are not to be merely tolerated in a democracy; rather they are the very lifeblood of democracy. Factions within a democracy, far from being primarily a threat to be controlled, as Madison (1987[1788]) thought, are not only inescapable but also necessary for democracy to flourish. In contrast to actually existing liberal democratic capitalism, "a healthy democratic process calls for a vibrant clash of political positions and an open conflict of interests. If such is missing, it can too easily be replaced by a confrontation between non-negotiable moral values and essentialist identities" (Mouffe 1993, p. 6). In that claim she is backed by Flyvbjerg (1998a, p. 209) who asserts that "social conflicts are the true pillars of democracy" (see also van den Brink 2005; Lefort 1988; Lefebvre 2003, p. 137; Connolly 1991; Wolin 1994).

However, Mouffe also offers a forceful critique of particular principles in liberal democracy (see also Levine 1981; Schmitt 1971; Pateman 1985; Eckersley 2004). Nancy Fraser (1990) has also been a leader in this charge, making clear from a feminist viewpoint the problems with the liberal conception of the public sphere. Liberal democracy tends to see the family and the home as largely private and outside the public sphere. As a result, liberal democracy effectively turns a blind eye to the patriarchal relations that govern families (see, among others, Jaggar 1988; Pateman 1987). The public/private distinction thus helps insulate patriarchy and misogyny from democratic challenge, just as it makes the world safe for the hierarchical social relations of capitalism, as other democratic traditions have pointed out. Moreover, Fraser argues that the liberal idea of the public sphere is flawed in several ways. [10] First, it imagines citizens to be free and equal agents who shed their particular identities and interests and come to reason together in public. That notion, that one can and must "bracket" one's particular positions in order to participate in the public sphere, is false, she argues. It reinscribes structural advantage for dominant groups, since the rules of the game are the specific values of white, western, and bourgeois males, which are instead thought of as "universal" human characteristics. "In stratified societies," she argues (Fraser 1990, p. 64), "unequally empowered social groups tend to develop unequally valued cultural styles. The result is the development of powerful informal pressures that marginalize the contributions of members of subordinated groups both in everyday life contexts and in official public spheres" (see also Mansbridge 1990). Iris Marion Young (1996, pp. 128–33) picks up this critique in more detail and applies it more specifically to deliberative democracy, arguing that discursive strategies that are more commonly used by marginalized groups, like greeting, rhetoric, and storytelling, sit at best uneasily with deliberative democracy's insistence on public reason-giving (see also Sanders 1997).

Perhaps the most dangerous part of the bracketing for Fraser is that it makes it seem as though it is unnecessary for a polity to eliminate social inequality in order to have democracy. We can simply cordon off that inequality in the private realm; in the public realm we can assert a formal political equality that is sufficient to meet the requirements of democracy. It is thus a central assumption of the liberal-democratic imagination that a democratic society can achieve real political equality even as it maintains material and cultural inequality. In that light, democracy and capitalism are compatible. For some, such as revolutionary democrats and, in fact, even de Tocqueville himself (2004, p. 60), that idea is a fantasy sure to crumble: "it is impossible to conceive of men as eternally unequal in one respect but equal in all others. Eventually, therefore, they will be equal in everything." For Fraser, however, de Tocqueville's "eventually" doesn't seem to have arrived, and in fact the idea of bracketing power helps ensure a lasting inequality outside the public sphere because it insulates that inequality from democratic challenge. Without explicit acknowledgment of difference and inequality, and a concomitant establishment of concrete ways of mitigating that inequality, structural inequalities will be preserved (Young 1990, 1999b; Yuval-Davis 1997, 1999).

A second critique that follows from the bracketing critique is that the liberal idea of the public sphere tends toward a unitary conception; it imagines the polity should have a single public sphere in which all should participate as equals. For Habermas, she argues (Fraser 1990, p. 62), "a single, comprehensive public sphere is always preferable to a nexus of multiple publics." However, for Fraser a unified public sphere enables the systematic disadvantage we saw in the previous critique. She counters that democracy is best served by multiple publics, by the proliferation of what she calls "subaltern counterpublics." Those are "parallel discursive arenas where members of subordinated social groups [women, workers, peoples of color, and gays and lesbians] invent and circulate counter-discourses, which in turn permit them to formulate oppositional interpretations of their identities, interests, and needs" (Fraser 1997, p. 81). Counterpublics "function as spaces of withdrawal and regroupment ... [and] as bases and training grounds for agitational activities directed toward wider publics" (p. 82). They are the incubators, in other words, for counter-hegemonic understandings of the world. Without such opportunities to interact, regroup, and organize, she believes, subordinate groups would routinely lose out in the unified liberal public sphere. In extending that argument to neoliberalization specifically, we might say that it is only *outside* the dominant public sphere, saturated with the logic of neoliberalism, that alternative political-economic visions can be nurtured and develop.

A last critique of liberalism focuses on its aggregative approach. Recall that both deliberative and participatory democrats reject the liberal tendency to discover the will of the people by aggregating individual preferences, a feat most commonly achieved through elections. Laclau and Mouffe (2000, p. xvii) stand with the other critiques on that point.

> Like them, we criticize the aggregative model of democracy, which reduces the democratic process to the expression of those interests and preferences which are registered in a vote aiming at selecting leaders who will carry out the chosen policies. Like them, we object that this is an impoverished conception of democratic politics, which does not acknowledge the way in which political identities are not pre-given but constituted and reconstituted through debate in the public sphere. Politics, we argue, does not consist in simply registering already existing interests, but plays a crucial role in shaping political subjects.

To be sure, not all liberal theorists offer so simple a vision of aggregative democracy. The target Mouffe and others are primarily aiming at here is the common-sense conception of democracy-as-elections that we met in the vignette about the Iraqi vote. Whatever the subtleties of liberal-democratic theory, the important critique here is of the way liberal democracy is being understood and actively portrayed—by neoliberals and by the media—

as an aggregation of the preferences of self-contained and autonomous individuals.

We can see in the quote above that Laclau and Mouffe conceive of power and the formation of political agency in a very particular way. As we saw, they see power relations and conflict as constitutive of social relations. "Power," Foucault argued (1988), "is always present." Political agents are "constituted and reconstituted" for Laclau and Mouffe through their conflictual relations with other members of the polity. Thus identities are not discrete and fixed, and we must renounce the liberal "category of the subject as a unitary, transparent, and sutured entity" if we are to understand politics at all (Laclau and Mouffe 1985, p. 166). Hence power and political conflict, because it is constitutive of social relations and therefore inescapably present in all polities, becomes the central focus of Mouffe's later work (1995, 2000, 2005). The problem of democratic politics, she argues, is not how to eliminate or tame power, as Habermas hoped, but how to "constitute forms of power [and conflict] that are compatible with democratic values" (1999, p. 753). Here she is following quite closely behind Foucault (1988), who argued "the problem is not of trying to dissolve [power] in the utopia of a perfectly transparent communication, but to give ... the rules of law, the techniques of management, and also the ethics ... which would allow these games of power to be played with a minimum of domination."

In order to begin to constitute power in this way, Mouffe makes a distinction between what she calls "antagonism" and "agonism." Each is a form of conflict, but antagonism refers to one group trying to eliminate the other from the polity; it implies a war of existence. Agonism, on the other hand, refers to groups consciously struggling against each other to gain hegemony, but each recognizes the other's right to exist. Agonism very explicitly retains an "us vs. them" approach to politics, but it rejects the war of existence that antagonism demands. Agonism

> presupposes that the "other" is no longer seen as an enemy to be destroyed, but as an "adversary," i.e., somebody with whose ideas we are going to struggle but whose right to defend those ideas we will not put into question ... An adversary is a legitimate enemy, an enemy with whom we have in common a shared adhesion to the ethico-political principles of democracy.
>
> (Mouffe 1999, p. 755)

Agonism and its category of the adversary do not eliminate antagonism from the polity, for antagonism is an inescapable feature of modern social life. Rather agonism seeks to engage antagonism, to channel it into agonistic relations. The goal of *democratic* politics, for Mouffe, is to transform antagonism into agonism. In insisting on the importance of an agonistic, adversarial approach to politics, Mouffe uses the term "the political." The term refers to "the dimension of antagonism that is inherent in all human society." That antagonism must be acknowledged, and then it can be managed. For Mouffe "politics" is the ensemble of practices, discourses, and institutions that can shape "the political" in democratic ways. In other words, in a healthy democracy "politics" is able to transform "the political" into agonistic struggle. Mouffe categorically rejects the notion that we can or should eliminate the political from the polity, as deliberative democracy seeks to do. It is hard to overemphasize the importance of "the political" to Mouffe's theory. Titles such as *The Return of the Political, On the Political,* and "Democracy, Power, and 'the Political'" make clear her insistence on its importance. Her fear is that political theory is evading and failing to apprehend "the political" (Mouffe 1993). At the heart of her critique of deliberative democracy is what she sees as that tradition's recidivist denial of "the political" as the critical foundation of democracy.

Mouffe's critique of certain elements of deliberative democracy (and, to a lesser extent, participatory democracy) is distinctly withering. She takes issue with two elements of deliberative democracy: (1) its mistrust of and desire to eliminate power and conflict, and (2) its predilection for consensus and the common good. Mouffe's opposition to the first element is clear in the discussion of agonism above. Recall Habermas' desire to have political questions decided by the forceless force of the better argument rather than by raw political power or force. Unlike the liberal desire to act as if power did not exist, Habermas' approach acknowledges existing power differences and seeks actively to neutralize them through communicative action, to ensure that power is not the driver of political decision-making. That goal parallels the perspective of the classic pluralists in assuming that power is discrete and alienable. That is, it conceives of an agent's power as a discrete resource that she possesses. Discrete power can be neutralized, set aside, contained, and the agent can go on operating without it. Agents can engage other members of the polity without exercising power. A relational view of power, on the other hand, holds that power inheres in the relationship between social agents (Laclau 1996, p. 27). Therefore, power is not an alienable quality that can be temporarily neutralized through skillful mediation. Rather it is ineradicable because it is constitutive of social relationships (Hillier 2003; Mouffe 2005, p. 18; Huxley 2000, p. 372; McGuirk 2001, p. 213). It is therefore always present and always impacting human relations. We cannot neutralize it any more than we can neutralize social relations themselves. Moreover, as Flyvbjerg (1998b), Hillier (2003), and Huxley (2002) point out, any active attempt to neutralize power through communicative facilitation is itself an imposition of a particular values and relations of power.

The problem with the ideal of a power-tamed deliberation, for Laclau and Mouffe (1985) and Mouffe (1993, 2002, 2005), in addition to its impossibility, is that it tends to diminish emphasis on "the political." If democracy's great strength is that it can embrace irreducibly conflictual social relations and channel them to productive ends, we must oppose a vision of democracy whose goal is to eliminate those relations. The deliberative ideal, in seeking ultimately to progressively minimize the us/them distinction, to emphasize "shared" interests, to constitute a comprehensive "we" (Mansbridge 1992), undermines the very strength of democracy. Such "suturing" of society's irreducible fissures, for Laclau and Mouffe (1985, drawing on Lacan), is both an impossible and an undesirable project (see also Connolly 1991). Democratic politics, in their view, should not be a search for intersubjective understanding and agreement. It should not attempt to overcome political disagreement and conflict. It should, rather, embrace disagreement and conflict. It should understand that politics is necessarily a struggle for hegemony among agents whose relations are irreducibly conflictual. The goal is not to develop *a priori* processes to control, neutralize, or eliminate conflictual relations of power. It is instead to *transform* those relations: to *mobilize* power to engage in counter-hegemonic struggles and to establish new hegemonies (Lefort 1988). Writing about the current neoliberal hegemony, Laclau and Mouffe (2000, pp. xvi–xvii) argue

> the present conjuncture, far from being the only natural or possible societal order, is the expression of a certain configuration of power relations. It is the result of hegemonic moves on the part of specific social forces which have been able to implement a profound transformation in the relations between capitalist corporations and the nation-states. This hegemony can be challenged. The Left should start elaborating a credible alternative to the neo-liberal order, instead of trying to manage it in a more humane way. This, of course, requires drawing new political frontiers and acknowledging that there cannot be a radical politics without the definition of an adversary. That is to say, it requires the acceptance of the irreducibility of antagonism.

In other words, an approach that confronts neoliberal hegemony with a cooperative search for a shared understanding and agreement cannot foster the kind of counter-hegemonic politics we require to challenge neoliberalization. "Instead of trying to erase the traces of power and exclusion," Mouffe (2000, pp. 33–4) argues, "democratic politics requires that we bring them to the fore, to make them visible so that they can enter the terrain of contestation." The power and exclusion that neoliberalization produces, in her vision, should not be excluded from politics, as deliberative democrats desire. It should, rather, "enter the terrain of contestation"; it should be the focal point of democratic politics and struggle.

Exacerbating the lack of "the political," at least in more faithful versions of deliberative democracy, is a commitment to the politics of the common good over and above the politics of particular interests. Whereas in strategic action participants pursue their particular interests over the interests of others, communicative action urges that participants reason together toward a shared understanding of the common good. Such a vision relies on the progressively greater "suturing" of society we saw above. Communicative action is designed to suture differences sufficiently to produce an agreement on what the common good entails (and, therefore, actions that will best realize that good). The suture metaphor is a powerful one, for it highlights the fact that the common-good approach sees difference as a *wound* that should be healed. A suture holds a wound together long enough for the body to heal itself. If, however, we see difference as an *orifice*, as a rupture whose very function is to remain open, then a suture (on, for example, the mouth, nose, or ears) imperils the body rather than healing it. If, like an orifice, social and political difference cannot and should not be sutured, if even small groups contain a level of fundamental antagonism that cannot be overcome through intersubjective understanding, then every seeming agreement is really always a temporary hegemony of some interests over others (Hillier 2003, p. 42). That would be even more true of every apparent consensus. If antagonism is irreducible, then every agreement will silence some and not others, and every decision will favor some over others (McGuirk 2001, p. 213; Tewdwr-Jones and Allmendinger 1998; Hillier 2002). Every agreement or

> consensus exists as a temporary result of a provisional hegemony, as a stabilization of power that … always entails some form of exclusion. The ideas that power could be dissolved through a rational debate and that legitimacy could be based on pure rationality are illusions which can endanger democratic institutions.
>
> (Mouffe 2000, p. 104)

"To negate the ineradicable character of antagonism and to aim at a universal rational consensus," she continues (p. 22), "this is the real threat to democracy" because it denies the critical democratic potential of "the political." It sutures an orifice that must remain open.

Agreement and consensus, therefore, do not exist as the result of the successful neutralization of power and the intersubjective discovery of a creative win–win solution, however much advocates wish it (and narrate it) to be so (see, for example, Susskind *et al.* 1999). Even if communicative advocates readily acknowledge the difficulty of achieving substantial agreement around the common good, the fact that they aim at it, that they conceive of it as possible and desirable, means that when participants successfully agree to a course of action, that outcome will very likely be accepted as a decision that is in the best interests of all. Such outcomes very often emphasize the achievement of intersubjective agreement and are at pains to take seriously the skillful practice used to achieve it (Forester 1998). They suggest that differences have been overcome, that conflicts have been resolved, at least on issues relevant to the present agreement. But Laclau and Mouffe contend that conflict can only ever be masked by agreement, not transcended, even in the short term. Such masking can be an extremely useful tool for neoliberal interests because existing power relations can remain unchallenged at their core even as significant political legitimacy is conferred (Fainstein 2000).

That critique of the common good is reinforced by a linguistic argument. Central to the deliberative and communicative project is the idea of undistorted communication whereby political actors progressively improve their ability to communicate such that they are able to work toward eliminating deception, coercion, and strategic action, and they increasingly communicate genuinely in an effort to reach intersubjective understanding. The counter-argument is that such undistorted communication is a logical impossibility, that distortion is a necessary element of language and communication. Mouffe (2000) draws on Wittgenstein to argue that actual communication is always shot through with partial understanding and communicative distortions. For Wittgenstein, if we actually achieved an ideal communication we would find "we have got on the slippery ice where there is no friction and so in a certain sense the conditions are ideal, but also, just because of that, we are unable to walk: so we need *friction*. Back to the rough ground" (Wittgenstein 1953, p. 46e). The notion of logical impossibility is picked up in Mouffe's (1999) and Hillier's (2003) analysis through Lacan. In everyday practice, they argue, all language can only ever *represent* the actual thing it aims to signify. There is an irreducible gap between signifier and signified (Hillier 2003). Therefore, a "field of consistent meaning," which we need in order to make sense of language, cannot be anchored in the actual things it seeks to represent (Mouffe 1999, p. 751). It must instead be anchored in something else, something chosen by the participants. Participants must therefore *impose* what Lacan calls a "master signifier," a crux that sets the relationships between signifiers and signifieds more generally and creates a consistent field of meaning. That master signifier necessarily *distorts* the symbolic field by elevating one particular representation over others (Zizek 1992, Chapter 3). However, the master signifier also holds the field of meaning together. It makes communication possible. Therefore, removing the distortion would cause the field to disintegrate, and communication would be impossible. According to that argument, distortion is therefore *necessary* to make communication possible. An ideal of undistorted communication is not merely difficult, it is oxymoronic. That fact is important because it means that language and communication, the centerpiece of the communicative approach, can never be a neutral, fully shared, and undistorted medium. Rather language and communication are always political; their distortions establish and reinforce hegemonic relations among participants. That realization leads Mouffe (2000, p. 98) to conclude that we should not be attempting to *eliminate* distortion and create power-free communication; rather we should acknowledge that distortion is always present and analyze who it favors and who it does not. Language and communication should, to reframe her quote above, "enter the terrain of contestation." Creating elaborate techniques to reduce distortion and power in communication can never neutralize or eliminate them. But deliberative techniques can tempt us to *think* we have, and so put us in danger of masking their operation.

Another way deliberative practices can mask what is taking place has to do with the Habermasian ideal of inclusiveness, whereby all affected parties must be involved in making a decision (Healey 1997). The problem with that ideal, radical pluralists argue, is that such inclusiveness can never be total. Every group that includes must always also, by definition, exclude. In order to exist, every "we" needs also to have a "they" (Hillier 2003, p. 42). That critique is rooted in Derrida's notion of the "constitutive outside," the idea, in part, that every identity must be constituted as much by what it is not (its outside) as what it is (its inside) (Mouffe 1993, p. 141; Connolly 1991; Derrida 1988). The constitutive outside is a *necessary* part of all social identities. Therefore, the ideal of inclusiveness must always go unrealized; every decision-making group must always exclude some affected parties in favor of others. Moreover, even if such an ideal were logically possible, it would be so practically difficult as to be virtually impossible. The logistical task of ensuring that *all affected parties* actually manifest as invested, empowered stakeholders is far beyond the resources (financial, imaginative, communicative) of any agency that is conducting a deliberative process. And then, on top of that, even if a truly inclusive body of stakeholders were assembled, the agency would have to achieve a power-free, fair, and undistorted deliberation. In practice of course, what happens is that agencies get the *most* affected stakeholders to the table (or, more accurately, they get *representatives* of the most affected stakeholders), and exclude relatively less affected stakeholders. While that method is practicable, it does not approach the communicative ideal. What is worse, the gap between reality and ideal is papered over far too easily in public discourse, so that processes that are necessarily exclusive get narrated as inclusive. In fact, all decisions taken through communicative action will be imposed on people who have not had a say in the process, but who are nevertheless affected by the decision. And that exclusion is all too often not random, but systematic. It may (or may not) be going too far to say that poor and non-white communities tend to be routinely excluded from such processes. However, it is not at all far-fetched to say that under the hegemony of neoliberalism, property owners and other large corporate interests will never be among those excluded. They are, therefore, systematically *included*, and therefore advantaged, by a decision-making structure that must of necessity exclude some affected parties, but never excludes *them*. Moreover, that process is commonly narrated to be inclusive, and its exclusions are rarely questioned.

Radical pluralism, on the contrary, assumes that exclusion is a normal part of politics. Its goal is not to eliminate exclusion, to imagine it has been averted, but to accept that it is unavoidable and to critically examine how we do it. It seeks to make transparent the question of who to include and who to exclude, to make it a central issue in political debate.

Yet another form of masking in deliberative democracy has to do with cultural esteem (Young 1996, 1999a). Habermas' theory of communicative action poses its mode of communication as universal, as common to all people regardless of culture, class, gender, etc. Because it relies so heavily on persuasion through rational argumentation, the deliberative ideal depends on "equality in epistemological authority," as Lynn Sanders (1997, p. 349) puts it. That is, it requires an equal "capacity to evoke acknowledgement of one's arguments." However, as we all know, some people "are more likely to be ... learned and practiced in making arguments that would be recognized by others as reasonable ones." As a result, some people are inevitably

> less likely than others to be listened to; even when their arguments are stated according to the conventions of reason, they are more likely to be disregarded. Although deliberators will always choose to disregard some arguments, when this disregard is systematically associated with the arguments made by those we know already to be systematically disadvantaged, we should at least reevaluate our assumptions about deliberation's democratic potential.
>
> (Sanders 1997, p. 349)

Sanders' argument is essentially a concern about the politics of "recognition," about the unequal esteem that different groups are granted by a dominant culture (Taylor 1992; Honneth 1995; see also Fraser 1995). In debates about social justice, many struggle over just how much recognition should be emphasized over and above questions of material inequality (Fraser 2001). However, there is wide agreement that cultural recognition is unequally distributed in society, and that such unequal distribution in many ways mirrors the unequal distribution of other resources. The debate about recognition runs directly to questions of gender, race, class, and sexuality. Patriarchies systematically devalue cultural traits considered not masculine; racist societies systematically devalue cultural traits considered not of the dominant race; etc. If people are not equally esteemed, their arguments will not carry equal weight in deliberation, and critics argue that is a problem even skilled facilitation can only weakly mitigate (Young 1999a; Sanders 1997). Dominant classes, genders, races, and sexualities begin with greater epistemological authority before they even open their mouths. Habermas' desire to remove physical force and coercion from democratic decision-making and replace them with "the force of the better argument" is admirable (Habermas 1985). But the unequal distribution of esteem means that the new "forceless force" will virtually always systematically advantage some over others (McGuirk 2001, p. 207; Burgess and Harrison 1998).

If we, as a result of those critiques, reject the search for intersubjective understanding and the pursuit of common-interest politics, we open the door to an entirely different politics. We can conceive of democratic politics as the mobilization of social movements that struggle for democratization. While many deliberative and participatory democrats reject social movements as undemocratic because they pursue particular, not general interests (e.g. Innes and Booher 2000), the radical pluralist approach embraces them. More than that, it finds such movements essential to a vibrant democracy. For Mouffe, diversity and pluralism, which social movements help invigorate, are fundamental to democracy. However, Mouffe's insistence on agonism means that members of the polity must adopt one overarching, though thin, consensus: a "shared commitment to the ethico-political principles of democracy." For her those ethico-political principles are liberty and equality. Members of democratic polity must share a commitment to those values. However, they should not be expected to agree on what those concepts mean.

> We can agree on the importance of "liberty and equality for all," while disagreeing sharply about their meaning and the way they should be implemented, with the different configurations of power relations that this implies. It is precisely this kind of disagreement which provides the stuff of democratic politics and it is what the struggle between left and right should be about. This is why, instead of relinquishing them as outdated, [the left] should redefine those categories.
>
> (Mouffe 2000, pp. 113–14)

But while Mouffe advocates a democratic approach rather than a social-revolutionary one, nevertheless she is committed to particular outcomes. Because every struggle necessarily results in a hegemony, she does not seek the elimination of hegemony, but the construction of a new one. Recall Laclau and Mouffe's argument (2000, pp. xvi–xvii) that the current neoliberal world is "the result of hegemonic moves on the part of specific social forces which have been able to implement a profound transformation in the relations between capitalist corporations and the nation-states." But, they go on, "this hegemony can be challenged."

Their political vision for realizing an alternative hegemony bears a strong resemblance to the way Hardt and Negri imagine the multitude. They understand that oppositional groups are scattered, and each cannot upend neoliberal hegemony alone. They therefore advocate a political strategy they call "chains of equivalence," whereby diverse movements come together to challenge neoliberalization. In a chain of equivalence, each movement retains its particular character, but it resolves to act in concert with other movements who occupy an *equivalent* position with respect to the neoliberal hegemony. As with Hardt and Negri's "multitude," Laclau and Mouffe spend much effort working through the Janus-face of their key concept of "equivalence." To be equivalent, two entities must simultaneously be both different and the same. In their words, there is an unavoidable "ambiguity penetrating every relation of equivalence: two terms, to be equivalent, must be different—otherwise there would be simple identity [i.e. they would be identical]. On the other hand, the equivalence exists only through subverting the differential character of those terms" (Laclau and Mouffe 1985, p. 128). Equivalence, therefore, always implies a paradox: sameness and difference held together in unavoidable tension. A chain of equivalence among different groups must simultaneously maintain their difference and articulate their shared goals. As with the multitude, Laclau and Mouffe do not imagine shared goals to exist *a priori* for the actors to discover. Rather their commonality is *produced* through conscious mobilization. The sameness in a chain of equivalence is not "the expression of a common underlying essence but the result of political construction and struggle" (Laclau and Mouffe 1985, p. 65). As with Hardt and Negri, they envision the creation of links among "the series of resistances to the transnational corporations' attempt to impose their power over the entire planet" (Laclau and Mouffe 2000, p. xix). Creating such links among diverse democratic struggles would produce a new hegemonic project to resist neoliberalization. "What is at stake," they conclude (2000, p. xix), "is the building of a new hegemony. So our motto is: 'Back to the hegemonic struggle.'"

A common, though to some extent inaccurate, criticism of Laclau and Mouffe is that while their critique is trenchant, they are less clear on just what their "new hegemony" would entail. In fact they do elaborate on this point. In sum, they imagine a new hegemony of radical democratization and equalization. I develop more fully just what I think that might mean in the next chapter, where I elaborate the democratic attitudes that can help us resist neoliberalization and imagine more democratic urban futures.

CHAPTER **3**

NEW DEMOCRATIC ATTITUDES

The whole history of the progress of human liberty shows that all concessions yet made to her august claims have been born of earnest struggle. The conflict has been exciting, agitating, all-absorbing, and for the time being, putting all other tumults to silence. It must do this or it does nothing. If there is no struggle there is no progress. Those who profess to favor freedom and yet depreciate agitation, are men who want crops without plowing up the ground, they want rain without thunder and lightening. They want the ocean without the awful roar of its many waters. This struggle may be a moral one, or it may be a physical one, and it may be both moral and physical, but it must be a struggle. Power concedes nothing without a demand. It never did and it never will.

—Frederick Douglass (1985[1857])

In Seattle in late Fall of 1999, as part of the wider protests against neoliberal globalization, members of the Direct Action Network successfully blocked key streets in order to prevent delegates from attending the 3rd Ministerial conference of the WTO. The next day, the City of Seattle declared a "no protest zone" around the convention center to prevent a repeat performance. Activists responded by breaching the perimeter of the zone. They eluded police for a time, but they were eventually apprehended and handcuffed. As they were being detained, they repeated a chant heard often throughout the week-long protests: "This is what democracy looks like." Democracy, they were saying, is not only voters in booths, or rational debate on the floor of the U.S. Congress, it is also shouted dissent; it is resistance; it is mass social movements; and it is political struggle in the streets of the city. It demands a right to be present in the city, to inhabit it, to occupy it, and to use it as a political forum. From their perspective, nothing is more anti-democratic than a no-protest zone in the heart of the city.

Chapter 2 makes clear that democracy can take many forms. The Seattle protesters suggest that some forms are dominant and others subordinate, and that subordinate forms must struggle continuously to imagine, articulate, and defend their vision of democracy. The dominant form of liberal democracy, Frederick Douglass implies, will concede nothing without a demand. This chapter presents a set of oppositional democratic attitudes to resist neoliberalization and imagine more democratic urban futures. It builds those attitudes out of the raw material of the traditions in Chapter 2, by aligning them either with or against particular logics in the various traditions. The first part of the chapter is devoted to articulating those democratic attitudes. The second part then develops a particular way of understanding democracy that is both spatial and urban. Most of the democratic theory in Chapter 2 lacks a spatial imagination. That is, democratic politics are imagined to take place in an abstract terrain. But as geographers have made clear, spatial relations are deeply and inescapably intertwined with political, social, and economic relations (Soja 1980; Massey 2005; Sack 1993; Thrift 1996). Any project to democratize cities must take account of the importance of democracy's spatial and urban dimensions. So the goal of the second part of the chapter is to take that into account. To do so, I draw inspiration from Henri Lefebvre's concept of the right to the city.

DEMOCRATIC ATTITUDES AGAINST NEOLIBERALIZATION

It is probably clear from the drift of Chapter 2 that my democratic attitudes rely greatly on radical pluralism. However, they are not identical to that tradition. They also draw on revolutionary, participatory, deliberative, and even liberal democracy. In Chapter 1, I make clear that a central goal of democratization must be to resist neoliberalization. Therefore, the first democratic attitude is to *(1) explicitly reject the current hegemonic pairing of neoliberalism and liberal democracy*. Recall that the argument here is that each needs the other, that you cannot have democracy without a capitalist economy (e.g. Friedman 1962). Central to my democratic attitudes is the need to reject that pairing, and in fact to actively undermine its currency wherever possible. The alternative is not that capitalism and democracy are entirely antithetical, but that they have no essential relationship. Therefore, it becomes entirely possible to imagine democracy, even liberal democracy, as *resistance* to neoliberal capitalism, rather than as its twin. As part of this rejection, we must also be extremely skeptical of the liberal tendency to erect barriers between public and private. More specifically, we must explicitly reject the liberal desire to imagine that the economy is part of the private sphere, and that economic relations are not properly subject to democratic politics. In order to resist neoliberalization's tendency to give capital more and more control over urban decisions, we must work to democratize corporations and other institutions that control capital. Here I stand with many deliberative, participatory, revolutionary, and radical pluralist democrats who have strongly opposed this liberal separation of economy and polity, and have called for more democracy in spheres outside the traditional public sphere.

While that first rejection is fairly straightforward (given my desire to resist neoliberalization), the second is a bit more provocative. Our democratic attitudes should *(2) reject the argument that the proper aim of democratic decision-making is to achieve consensus and/or the common good.* Here I part company significantly with deliberative and participatory democrats, and in a different way with most revolutionary democrats. I adopt a more Foucauldian (or more specifically Mouffian) political approach instead of a Habermasian one. I argue that the ethic of the common good poses problems for democracy in general, and it poses important democratic problems specifically in the context of neoliberalization.

In general, radical pluralism's critique makes a strong case that the common good approach tends to perpetuate inequality. Even when participants come close to the lofty ideal of communicative action—that is, they do not act strategically in their self-interest, but collaboratively in the common interest— inequality is nevertheless preserved. Given pre-existing inequalities, the demand that all groups put the common good ahead of their own interests imposes an unequal burden. For example, it is a much greater demand to ask a poor neighborhood to eschew its particular interests than it is to ask the same of a rich neighborhood. The common-good ethic restricts the political options of disadvantaged and marginalized groups. The most effective way for them to overcome their disadvantage is to organize and advocate for their *own* interests. A social-movement model where disadvantaged groups come together to pursue democratic outcomes that best meet their particular interests, such as that advocated by Laclau and Mouffe, is the political option that can most directly redress inequality. The common-good ethic stifles such an option; it would require the social movement not only to achieve the decisions that favor them, but also to do so by arguing that those decisions favor not only them, but the more general interests of the city (or the society) as a whole. Such a burden is almost punitive. As we will see in Chapter 4, South Park is a relatively poor neighborhood in Seattle with a disproportionate percentage of residents of color. It has been subjected to systematic environmental injustice over the past forty or so years. Imagine they mobilize to resist the citing of yet another hazardous waste facility in their area.[1] To their claim that such a facility would negatively impact the well-being of South Park residents, they are told by communicative facilitators to "leave their pet pigs at the door," and encouraged to make their argument in terms of what is best for the city as whole. Even if they played the game and tried to make a common-good case, it is very likely that technical experts would be waiting with a well-developed counter-argument, based in quantitative empirical data, that for reasons like efficiency, links to transportation lines, and cost of land, the South Park site was in fact the best for the city as a whole. Even if it were possible to forge decisions that were equally in the interests of all, the common-good approach would thus limit the political horizons of less-privileged groups.

However, if we accept Mouffe's argument that all decisions are always in the interests of some more than others, then common-good approaches pose important ideological dangers as well. Any claim that a decision has achieved a common good, even if that claim were generated through deliberation by the participants themselves, covers up the unavoidably unequal outcomes of the decision. The decision cannot benefit everyone equally, even if it claims to. Therefore, as Mouffe argues, instead of aiming at a common good we should acknowledge the unavoidably conflictual character of politics and the hegemonic nature of political decisions. Rather than legitimating decisions by claiming they serve the best interests of everyone, we should be transparent about the fact that each decision is what Mouffe (1996, p. 10) calls a "temporary result of a provisional hegemony" that "always entails some form of exclusion." Such transparency is vital to democratic politics, and it is most vital to disadvantaged groups that have been on the short end of so many decisions.

The need for transparency is intensified when we consider the particular context of neoliberalization. As we saw above, while neoliberalization may produce significant democratic deficits, that does not mean its advocates abandon the rhetorical terrain of democracy. On the contrary, neoliberals actively co-opt democratic procedures as a way to legitimate their agenda. While liberal democracy is the preferred ally, as deliberative democracy has emerged as a popular alternative to liberal democracy, it has been similarly subject to co-optation (Swyngedouw et al. 2002, p. 209). And in fact it offers just what neoliberalization requires. Deliberative democracy's insistence on pursuing the common good tends to perpetuate inequality by providing weakened or non-existent means to challenge it (Fainstein 2000). At the same time, it usually produces a high degree of democratic legitimacy. Decisions made by consensus and in the common interest are seen as unassailably democratic. Deliberative practices therefore offer an extremely attractive partner for neoliberalization. It offers democratic legitimacy without significantly threatening existing power relations. In Chapter 4, I develop in detail how that process looks on the ground in my examination of the Seattle Waterfront.

Instead of eschewing or stifling deliberative democratic forums, business leaders and neoliberal policy-makers can choose to support and participate in them. In their ideal form, where all participants are selflessly pursuing the common good, the current hegemony of neoliberal common-sense looms large. Recall the insights of Larner (2000) and Giroux (2004), among others, who argue that neoliberalism has been highly successful in installing its particular vision as hegemonic common sense. The keystone of the neoliberals' success has been to establish their hegemonic ideological assumption: neoliberal policies are the best (and even only) way to ensure the common good. Pauline McGuirk (2001) observed just such a dynamic when she studied a deliberative planning process in Newcastle, New South Wales, Australia. The process operated under the assumption that Newcastle was a declining industrial city that desperately needed to enhance its economic competitiveness. Business interests used that narrative to equate their particular interests (maximize the exchange value of downtown property) with the interests of the city as a whole. Alternative interests, such as the use-values of residents, were unable to challenge the basic assumption: the city cannot afford to do anything other than what is best for business. Given this ideological hegemony, a democratic strategy that insists on framing arguments in terms of the common good is ripe for neoliberals to use to their advantage. We must reject the common-good approach and adopt instead Mouffe's view that *all* political projects are the pursuit of particular interests, that all claims to the common interest are necessarily false (even if they are sincere). We can then see neoliberalization, with Harvey (2005), as an agenda to restore the power of the capitalist class. Such transparency would help immeasurably to articulate the political interests that lie behind any assertion of the common good. And it would therefore foster far more effective resistance to the neoliberal agenda.

So far, my argument against the common-good approach has assumed that participants will actually *do* what communicative action requires. But another danger of the approach is that its ideal is terribly difficult for participants to achieve. Very frequently, they do not adhere to the Habermasian ideal of communicative action and instead pursue their self-interest. Consensus-building practices, for example, increasingly abandon the common-good ideal and acknowledge that "stakeholders enter the process to serve their interests. They give up nothing they have outside the process unless it benefits them" (Innes 2004, p. 14). Even so, such models do retain some other Habermasian elements. Innes' model, which is a fairly representative example of the consensus-building approach, preserves both the ideal of undistorted communication and the desire to neutralize power. She thinks (2004, p. 11) that "the technology is … very well developed on how to create undistorted communication or ideal speech situations." She also asserts (2004, p. 12) that power differences can largely be equalized for the purposes of deliberation "with skillful management of dialogue, shared information, and education of the stakeholders." So the model becomes one in which various stakeholders engage in communication undistorted by power to produce a creative new solution that is perceived by each group to benefit their self-interest. The value added here is that through deliberation participants invent new solutions they had not considered before coming to the table. Innes is unequivocal that in order to come to a shared solution, all participants must be satisfied with the outcome. No participant can believe their interests were not well-served. But by logical necessity, such a model guarantees that pre-existing relations of power will be reinscribed. Relatively more powerful groups can ensure their interests are met—it is, in fact, a requirement—and so there is no possibility of fundamentally transforming existing relations of power. Innes reinforces that point. While she believes that *at the table* power differences can be equalized by facilitators, she admits that *pre-existing* power relations are "untouched by consensus building … Consensus building is not, in any case, the place for redistributing power" (2004, p. 12).

That admission is striking. It makes Innes' consensus building entirely safe for neoliberalization, since it guarantees that the hegemonic position of capital cannot be significantly challenged. Moreover, it offers business interests an extremely attractive legitimation tool. And Innes doesn't hide this legitimating potential. The democratic deficits that neoliberalization produces means that business groups want the buy-in of "disadvantaged and minority stakeholders" in order to legitimate the decision (2004, p. 11). In cases of land development, for example, such legitimation problems can stall the process, and if they become chronic and generalized they can stall future development. In Innes' process, if developers incorporate the concerns of weaker groups into the agreed plan, "they can still get what they want without compromising their welfare" (p. 13). As long as they can ensure that the development goes forward (p. 15) in the short term and that development in general can proceed in a timely manner, developers' essential needs are met, and their projects are legitimated by the buy-in of disadvantaged groups. Such a model, far more nakedly than a more traditional common-good-seeking one, preserves and legitimates the status quo and guarantees that what Leonie Sandercock (1998, p. 176) calls the "structural transformation of … systematic inequalities" cannot take place.

If we reject the common-good model, it allows us to instead *(3) embrace an agonistic, social-movement model for democracy*. Deliberative democrats and communicative planners tend to see social movements as not fully democratic because they advocate for particular interests and exclude those not associated with the cause. And movements tend to be impassioned rather than cool and rational. However, from the point of view of Mouffe's agonistic pluralism, social mobilization is the very stuff of democracy. Mobilizations arise from the antagonisms that are always potentially present in society. Since antagonism is endemic to a differentiated polity, normal social life should produce a range of different movements that mobilize to both constitute and advocate for their interests. Laclau and Mouffe (1985) see mobilization as the process whereby groups produce, through struggle, an understanding of oppression. That is, they move from a relation of subordination, where a group is subject to the decisions of a different group, to a relation of oppression, where the subordinate group comes to construct its subordination to other groups as unjust and in need of remedy. From a radical pluralist perspective, therefore, social movements should not be thought of as groups "waking up" to their objective interests and mobilizing to achieve them, as with Marx's class *in* itself becoming a class *for* itself (Marx 1994, p. 182). Rather social movements are best understood as groups interpreting subordination as oppression, articulating antagonistic relations between themselves and other groups. Democratic social movements, in this vision, would then engage those other groups in agonistic struggle. As long as that struggle is agonistic and not antagonistic, as long as it does not become a war of existence, such social movements deepen democracy because they mobilize a proliferation of different cross-cutting perspectives and interests that engage each other politically. Multiple social mobilizations in pursuit of particular interests are, from this perspective, characteristic of a vibrant democratic polity. In Frederick Douglass' words, they "plow up the ground" to grow the crop of democracy.

That social-movement vision overcomes the common-good model's tendency toward unity and consensus, but it does not yet provide a way to redress the inequality that is produced by neoliberalization and reinscribed by common-good models of democracy. Multiple, disconnected social movements, pursuing all sorts of particular interests, are not likely to unseat an overarching hegemony like neoliberalism. What is required, therefore, is to *(4) embrace coalitions of movements.* For Laclau and Mouffe, this would involve their notion of "chains of equivalence," whereby a variety of movements remain distinct but construct a sense of themselves as sharing a relation of equivalence vis-à-vis some other social force. The Seattle demonstrations against the WTO, as we have seen, involved a range of distinct movements that came together to express an equivalent opposition to neoliberal globalization. They opposed it for different reasons (it lowers wages, it immiserates the developing-world poor, it turns a blind eye to human rights, it kills sea turtles), but they joined in a shared opposition to its broad agenda. Hardt and Negri (2004) imagine a very similar kind of politics. The "multitude" is irreducibly diverse, yet they actively produce and act upon their common position with respect to the dominant global political-economic system. Hardt and Negri imagine networks of movements that through their social and political actions produce "*the common* that allows [the multitude] to communicate and act together" (2004, p. xv). Their "common," while it is linguistically close to the terms used by deliberative and participatory democrats, is nevertheless closer to Laclau and Mouffe's concept of equivalence: it evokes a relation that is simultaneously (and paradoxically) different and the same.

As a metaphor, Hardt and Negri's image of a "network" among various movements is probably more descriptive of what is currently being practiced than is Laclau and Mouffe's "chain." However, Laclau and Mouffe's "equivalence" is a much more explicit construction of a relation that is simultaneously and paradoxically the same and different. Hardt and Negri's "common" is less explicitly designed to capture that relation. While it is true that Hardt and Negri clearly affirm the irreducible difference of the multitude, one gets the feeling their "common" is more firmly a commitment to unity that admits and wants to accommodate the unavoidable presence of difference. They seem less willing to *foster* difference than Laclau and Mouffe. Moreover, Hardt and Negri's language is a bad fit for the attitudes I have laid out, since the word is so redolent of a Habermasian politics of the common good. Hardt and Negri are quite different from Habermas, of course, but better to avoid the linguistic similarity and adopt "equivalence" instead. As a result, the democratic attitudes I advocate desire to produce what are best described as *networks of equivalence* among democratic social movements.

That way to imagine social-movement politics, while fairly new, is waxing. Most prominent are initiatives that are developing a working model for networked movements. Some few examples include the movement against neoliberal globalization, the Zapatista movement in Mexico, the Landless Workers' Movement in Brazil, the Narmada Valley dam protests in India, the meetings at the World Social Forum, the movement against GMO food, and blogging networks designed to provide alternatives to corporate media. The historical and political project here, as it is with Laclau and Mouffe and Hardt and Negri, is to overcome a well-recognized Catch-22: we want to avoid the reductionism of the Old Left that reduced everyone to their class position, but we also want to avoid the extreme fragmentation of radical postmodernism (Mouffe 1999). The goal is for difference to remain irreducible, for movements to remain fundamentally distinct from each other, yet to enable them to act in concert, to produce a coordinated opposition to their equivalent (not identical) oppression. An entirely reductive movement is not democratic (and could easily be totalitarian), but a plethora of disconnected movements poses no threat (and even offers much opportunity) to neoliberalization. Neither I nor the authors above purport to have the perfect *solution* to this Catch-22; that must be pursued through political struggle, struggle that is already underway. Rather I am arguing that a democratic resistance to neoliberalization should adopt an attitude that *aims to* construct viable networks of equivalence. In the case studies in Chapter 4, I develop some ideas about how we might perhaps achieve that difficult trick.

As I have sketched them so far, networks of equivalence do not necessarily pose an explicit challenge to neoliberalization. They could be developed around multiple agendas of equivalence, which may or may not oppose neoliberalization.

Therefore, we should, drawing from revolutionary democracy's opposition to capitalism, develop *(5) a clear commitment to oppose neoliberalization*. Because networks of equivalence adopt an agonistic approach, they would understand neoliberalization as a project to intensify the power of the capitalist class, and they would engage in agonistic struggle to undermine the neoliberal hegemony and forge an alternative one. The movements involved in such a network would conceive of themselves as sharing an equivalent position as not-capitalist-class and therefore disadvantaged by neoliberalization. However, the particular way each is disadvantaged would differ significantly, and those differences would certainly pose political challenges to the network's commitment to act in concert. However, an agonistic commitment to radical pluralism would ensure the network avoided the dangers of totalitarianism that haunt both the democratic models like that of Lummis, Rousseau, and civic republicanism, and the historical record of communist states in the twentieth century. Recall that radical pluralism's opposition to totalitarian visions is rooted in the soil of liberal democracy, which is to say the original liberal democracy Mouffe wants to rescue from the impoverished, actually-existing version that is neoliberalism's partner in crime.

That stance in opposition to neoliberalization is important both for avoiding cooptation to the neoliberal agenda, and for helping to constitute the equivalence that will bring together networks of movements. In forging equivalence and a concomitant political identity for a network, it is always valuable to have Derrida's "constitutive outside," an antithesis that the movement stands against. However, it is also necessary to advocate for a positive agenda as well as a negative one. Networks of equivalence must not only articulate what they are against, but what they are for. To that end, the counter-hegemony that these networks would struggle for to replace neoliberalization should be *(6) the counter-hegemony of radical democratization and radical equalization*. Radical democratization lies at the heart of the agenda of radical democracy. It demands a radical extension and deepening of democratic relations; it is an agenda to push democracy into spheres from which it has traditionally been excluded. As we have seen, the liberal-democratic model limits democracy to a narrowly defined public sphere, a sphere more or less coterminous with the state. It therefore leaves whole swaths of society unaffected by democracy's logic and values. Radical democratization's agenda is to overrun those limits, to radically extend democratic relations. Though he would not characterize himself as a radical democrat, John Dryzek (1996, pp. 5–7) talks about "deepening democracy" in a way that is quite compatible with a vision of radical democratization. He calls for advancement on three fronts. The first is what he calls franchise: the proportion of a polity's people that can participate in decisions. Here he has in mind the historical extension of democratic rights to non-property owners, women, non-white people, and the like. The second front is "scope": the domains of life that are subject to democratic relations. Scope concerns the extension of democracy to spheres like the economy and the home. That extension may or may not mean the expansion of the state relation. One might very well imagine the radical democratization of the corporation or the family without imagining the state absorbing those institutions into its formal structures. Dryzek's last front is "authenticity," by which he means "the degree to which democratic control is substantive rather than symbolic, informed rather than ignorant, and competently engaged." For example, pageants like the typical public hearing that appear democratic but don't take seriously the comments of participants are not "authentic" in Dryzek's sense.

Perhaps the most attention has been paid to the issue of what Dryzek calls scope. Democratization entails extending democratic relations into new spheres. Many focus on the economy, and more specifically the capitalist firm, with its traditionally hierarchical and authoritarian relations (Dahl 1985). For feminists, the key frontier is the home, the family, and the private sphere more generally (Marston 2000; Pateman 1987; Mansbridge 1990). One could imagine, as well, a host of other social spheres that lie beyond the traditional ambit liberal democracy assigns itself. Those might include, for example, the environment, science and scientific knowledge, media and information, genetic knowledge and technology, artistic creativity, and the like. Democratizing such spheres would first involve resisting their marketization and privatization, which is the ongoing agenda of neoliberalization. Democratization would instead involve a sense of a public, collective, and democratic approach to social relations in these spheres. It is important to be clear that the agenda here is *radical* democratization, not *total* democratization. One could imagine each of Dryzek's criteria being taken all the way to its endpoint such that *all* spheres of life are subject to *entirely* authentic democratic relations involving *all* people. Radical democratization, instead, advocates the radical extension of democratic relations on these fronts. It is an agenda to radically democratize the polity. We should expect that agenda to be vigorously contested. Radical democratization, therefore, does not expect or demand to arrive at a final end-state of absolute democracy. It aims to radically democratize our current political economy along the axes of franchise, scope, and authenticity, and it expects that process to be an ongoing struggle rather than a quick march to a crystal palace.

Just to give a small but perhaps key example of radical democratization from the world of information technology, it would oppose proprietary software developed by corporations and advocate the progressive hegemony of open-source programs that were produced and refined collectively by the universe of users. The production, distribution, refinement, and use of that technology would be a public and collective process. The rising presence of the open-source internet browser Mozilla Firefox (as opposed to Microsoft's Explorer) suggests that such a counter-hegemony is possible (Vaughan-Nichols 2006). For urban politics, radical democratization would involve resisting the marketization and privatization of urban space, and see urban development generally as a public and collective process involving as wide a range of urban inhabitants as possible. The anti-essentialist approach to radical democracy resists dictating the specific form such democracy should take. Rather it proposes an agenda of radical democratization that is likely to unfold in different ways at different times and places. Moreover, the agenda of radical democratization is ambivalent toward the state. It does not imagine democratization to be the same thing as the progressive extension of the formal sphere of the state. Because radical pluralists (as well as participatory and revolutionary democrats) reject the notion that democracy and politics are contained inside the state, democratization is by no means imagined to be the same thing as "stateization." So, while the state (significantly reimagined, to be sure) may well play a role in establishing a counter-hegemony (see, for example, Miller 2007, p. 228), it is not a necessarily central element of the agenda. That fact belies some anarchist critiques of radical pluralism (e.g. Day 2005), which assume, I think inaccurately, that a counter-hegemony approach like Laclau and Mouffe's hinges on an agenda to capture the state. So in sum, there is an element of uncertainty in the way forward. Outcomes of a democratic politics must by definition be shaped by the *demos* that engages in them. But that openness should not be mistaken for a lack of a clear vision: radical democratization envisions a fundamentally more public, more collective, and more democratic world than either neoliberalization or the liberal-democratic state can provide.

A necessary corollary to radical democratization is radical equalization. That agenda is potentially problematic, and it requires us to re-emphasize the commitment of radical democrats to something entirely different than the totalitarian outcomes of many socialist and communist states. Laclau and Mouffe, and to a lesser extent Hardt and Negri, explicitly reject the reduction of subjects to an *identical* political subjectivity because it stifles the difference that is fundamental to a democratic polity. The *equivalence* they insist on emphasizes the difference, in addition to the commonality, among political subjects. Radical equalization, then, in the context of radical pluralism, cannot entail *total* equalization whereby all people must have identical material circumstances or identical political power. It cannot mean people are reduced to being identical political subjects. Rather, it claims a *radical* equalization of material wealth, political power, and cultural esteem. It demands more than just the modest redistributions of the Keynesian era, whereby relatively poorer people (white male workers, primarily), won limited, though meaningful, material gains. It demands instead a substantial equality whereby all people are materially, politically, and culturally *equivalent* in a way that makes political equality truly possible. Nancy Fraser (1990) insists that democracy requires "substantive social equality." Her critical analysis of liberal democracy argues that the public sphere cannot bracket social inequality and hope to attain meaningful political equality. Democracy requires, instead, what she calls "participatory parity," which entails "social arrangements that permit all (adult) members of society to interact with one another as peers" (Fraser 2001, p. 29). For Fraser, democracy requires that no group be prevented from participating as a true peer in discussion and deliberation. A group cannot be a true peer, a different but fully respected partner, to other groups if it does not have equivalent resources, power, or cultural esteem. Again, that does not mean groups have to possess *identical* resources, power, or cultural esteem. They come by and mobilize those assets in different ways. But in order to be each other's peers, subjects must all possess those assets equivalently. In Fraser's view, then, radical equalization is a necessary part of radical democratization. So, in addition to *extending* democracy into spheres where it has not traditionally been practiced, we must also *deepen* democracy by pursuing radical equalization. Through equalization, we can come ever closer to the ideal of participatory parity, whereby all citizens really do participate as peers. Liberal democracy claims it offers such parity, but it falls woefully short of actually achieving it. Part of the agenda of radical democratization, then, is to make good on liberal democracy's false promise.

And so the networks of equivalence I advocate would resist neoliberalization and instead pursue a counter-hegemonic project of radical democratization and equalization. It is worth stressing that everything in that sentence is conceived of as a process. That is, the democratic attitudes resist neoliberal*ization* through radical democrat*ization* and equal*ization*. The end states of those processes may or may not be possible or, even if they were possible, desirable. It is likely that neoliberals would regret achieving a total neoliberalism because, if the work in regulation theory is any guide, it would collapse, and quickly, under its own weight. The same is likely true of both democratization and equalization (Lefebvre 1991a, p. 182). It is likely and probably desirable that the contingent political struggle for radical democratization and equalization will result in something quite a lot different than absolute democracy and total equality. And those are really not the goal of a counter-hegemonic movement. Rather the goal is to undermine the current hegemony, in which we constantly strive for a more neoliberal city, and to establish instead a different hegemony, where we constantly strive for a more democratic and more equal city. Hegemonies, by their nature, are always partial and always subject to challenge. We must never expect a total or eternal end state but rather a constant struggle to both subvert the dominant assumptions and values and to establish new ones. That struggle, for Laclau and Mouffe and for my democratic attitudes, is what democracy looks like.

A last commitment is an important strategic opening for the counter-hegemonic project I advocate. If democratic politics is a struggle for hegemony among competing interests, then it is important to take seriously all viable strategic options. One extremely important strategy in democratic politics has long been the discourse of rights. My claim is that democratic attitudes *(7) should be willing to engage the democratic discourse of rights.* It is not uncommon for democrats in both the participatory and especially the revolutionary tradition to be deeply skeptical of claiming rights as a political strategy (Tushnet 1984). The skeptics fear that because neoliberalism has so tightly tied itself to political liberalism, and because a central value of seminal neoliberal thinkers was to champion the rights of the individual against what they saw as the arbitrary power of the Keynesian state, that a rights strategy is doomed to be trapped within the neoliberal, or at least the actually-existing liberal, frame (Chandler 2002). "Neoliberal concern for the individual," many worry, "trumps any social democratic concern for equality, democracy, and social solidarities" (that is Harvey 2005 representing the concerns of others, p. 176). A democratic state's decision to pursue greater equality, for example, can be cast by neoliberals as tyranny, because it would compel wealthier individuals (including corporations) to submit to some form of redistribution, and violation of their property rights, presumably against their will.[2]

However, the liberal conception of rights is a distinctly individualist one, and that individualism is one element of actually-existing liberalism that Mouffe's agonism strives to overcome. The individualist notion of rights is rooted in the negative concept of freedom: individuals should not be *prevented* from doing as they wish (speaking, believing, being present) by an arbitrary power. That notion of rights sees rights as inherent, as vested in the individual. They are trumps designed to protect the individual against tyranny (see Dworkin 1984 for an exploration of trumps). However, rights can be seen in a very different way. They can, instead, be seen as collective political *claims* that groups make on a wider society. It is entirely possible to imagine rights as collectively asserted, as claims that a group of people make in the course of a democratic social movement to transform power relations (Sassen 1999). They might claim the right to clean water, or social insurance, or affordable medical care, or healthy food. A "claim" approach sees rights as associated with positive freedoms: people claiming a right to the basic necessities that enable them to thrive in the way they choose, that maximize their freedom to make their life on their terms. In that sense, as claims, rights have a long tradition in democratic struggle. Bowles and Gintis (1986) characterize the development of liberal democracy as a struggle between two conflicting trends: the expansion of property rights and the expansion of "personal" rights. The expansion of property rights has been coextensive with the expansion of capitalist relations of production, privatization, and marketization, and neoliberalization represents a recent and strong surge in that project. The expansion of personal rights, for Bowles and Gintis, stands against the expansion of property rights. That expansion has been won by struggle on the part of disadvantaged groups over the course of several hundred years. The progressive extension of suffrage, the development of workers' rights, and the enlargement of the social rights of welfare are all examples of rights that have been carved out of the flesh of property rights. Though Bowles and Gintis call them "personal," they are really more accurately collective group claims that have been made on the wider society. Similar rights are currently being claimed by social movements, such as a right to affordable housing, to good jobs, to safe and efficient public transportation. We might also imagine rights claims that strike even more at the heart of property rights, as with a demand that people participate meaningfully in the investment and management decisions of corporations.

In that light, property rights, or the right of corporations to profit, or their right to control investment, are merely claims, not trumps. They must be continually re-claimed and re-won through political struggle. Neoliberalization can be seen thus as a capitalist-class social movement that has had great success claiming and codifying property rights and the right to profit (Piven and Cloward 2000; Harvey 2003; see also Sites 2007). One effective response to neoliberalization is therefore a counter-movement, a network of equivalence that claims a different bundle of rights, rights that oppose that of property and aim at a more democratic and civilized polity (Harvey 2003). Given the demonstrated potential of rights discourse to win important political concessions, I think we should heed David Harvey's (2005, p. 179) admonition that it would be wrong to "abandon the field of rights to neoliberal hegemony." Harvey's position is supported as well by Bowles and Gintis (1986), Saskïa Sassen (1999), Leone Sandercock (1998), and Don Mitchell (2003), among others. History has shown that there are important battles to be fought and won on this field. One can imagine, for example, rights claims that extend democracy to new spheres or deepen it through social equalization. Among the many promising rights-claims against neoliberalization currently being pursued, in the remainder of the chapter I focus on one in particular: a claim to a "right to the city." I argue that claiming a right to the city has great potential as a way to confront the current political economy. In addition, it also offers us a way to imagine a democracy that is profoundly urban and spatial.

A Democracy that is Both Urban and Spatial

Although the democratic attitudes I have presented offer what I hope is a useful critical reimagination of existing democratic theory, so far they are neither particularly urban, nor particularly spatial. Both of these qualities, I argue, are important for the project of resisting neoliberalization. The urban component, for its part, is important for many reasons. The first is specific to the book: its particular topic concerns neoliberalization and democratic resistance *in cities*, and so our democratic attitudes should be explicit about what role the urban context plays in the larger political project. The second reason is strategic and tied to the historical development of both capitalism and neoliberalization. In general, capitalism tends to be closely tied to urbanization. Historically, capitalism has helped intensify processes of urbanization as industrialization made great concentrations of people not only possible but economically desirable (Castells 1977; Harvey 1985). As a consequence, resistance to capitalism has long had an urban character, as the clustering of workers near factories helped hasten the development of class consciousness and political activism. More specifically, cities have been key cites for the advancement of the neoliberal agenda. David Harvey argues this most strongly when he contends that the 1975 fiscal crisis, bankruptcy, and bail-out of New York City, in which financial interests were able to radically reorient the priorities of city government away from social provision and toward fiscal discipline,

> pioneered the way for neoliberal practices … it established the principle that in the event of a conflict between the integrity of financial institutions … and the well-being of citizens … the former was to be privileged. It emphasized that the role of government was to create a good business climate rather than look to the needs and well-being of the population at large.
>
> (Harvey 2005, p. 48)

Harvey concludes, with William Tabb (1982), that the subsequent rise of neoliberalism under Reagan and Thatcher was more or less the New York experience writ large. Cities, in a way, gave birth to neoliberalization. In addition, cities have been key laboratories in which neoliberal policies have undergone the process of trial and error (Brenner and Theodore 2002; Peck and Tickell 2002b). In his study of gentrification, for example, Neil Smith (1996, 2003) argues that cities, always important frontiers for capitalism's restless search for new profitable spaces, are growing increasingly so under neoliberalism. Given the importance of cities to neoliberalization, then, it would follow that we should take seriously the role that the urban might play in any democratic resistance.[3] One way that might happen, as the following discussion of the right to the city will suggest, is that different movements in the city can come together around shared claims to urban life (Castells 1983). Movements can see their particular struggles as part of a shared struggle for a different kind of city (Samara 2007). The urban can therefore be a strategic linchpin that holds together networks of equivalence.

The spatial component is vital as well. Because cities are themselves differentiated spaces and embedded in spatial networks, any project to resist the urban future offered by neoliberalization and advocate alternatives to it is already a spatial project. On top of that, if we heed the arguments of classic work in geographical political economy, we should understand neoliberalism as a necessarily spatial project (Harvey 1982; Lefebvre 1991c; Soja 1989; Smith 1984). In brief, the argument is that under capitalism spatial clustering produces both economies and diseconomies. The economies encourage further clustering, as when a concentration of software firms attracts skilled workers, which attracts still more firms, etc. The diseconomies encourage dispersal and relocation, as when concentration produces problems for capital like high land prices, taxes, wages, and transportation congestion. Such waxing and waning profitability means capital must restlessly engage in creative destruction, disinvesting in an area with low profitability and reinvesting in areas that promise a greater return. In the postwar era, both the large-firm mode of production based on economies of scale and the Keynesian economic policy ensemble tended to dissuade mobility, cloistering capital inside national borders. But capital mobility has always been integral to profitability, and the cement shoes of the Keynesian accommodation resulted in declining rates of profit. Understood that way, neoliberalism appears as a fervent, almost desperate attempt to win for capital more freedom to do what it *must* do: move. Neoliberalization, therefore, is a deeply spatial project. It follows that we must forge a democratic resistance and alternative to it that is also explicitly spatial (Lefebvre 1991c). Our democratic attitudes must simultaneously be political and geographical; they must reimagine politics and power relations, but at the same time they must also reimagine the meaning and purpose of urban space.

In the next section I make the case that the concept of the right to the city, if we construe it in a particular way, offers what we need. It allows us to infuse our democratic attitudes with both an urban and spatial understanding of democracy, an understanding that stands in stark opposition to the neoliberal agenda.

A Right to the City

Over the past several years, the idea of a "right to the city" has become increasingly popular. Academics have begun taking up the concept (e.g. Friedmann 1995; Isin 2000; Soja 2000; Pincetl 1994; Harvey 2003). Academic conferences have been devoted to the idea (Rights to the City 1998, 2002). In a recent volume, Don Mitchell (2003) collected a range of previously published work under its rubric. Many in that literature are exploring resistance to neoliberalization specifically (Sassen 2000; Holston 1998; Smith 1993; Salmon 2001). The idea has also become popular outside academia. To name just a few examples, it is being evoked in conflicts over housing (Grant Building Tenants Association 2001; Olds 1998) against patriarchal cities (City & Shelter *et al.*, no date; United Nations Center for Human Settlements 2001), for participatory planning (Daniel 2001), and against social exclusion in cities more generally (Buroni 1998; Cities for Human Rights 1998; Worldwide Conference on the

Right to Cities Free from Discrimination and Inequality 2002). Recently, a working group of the United Nations, under the joint purview of UNESCO and UN-HABITAT, has begun exploring how the right to the city could be used to pursue greater sustainability and social justice in cities. The group's goal is both to build ties among innovative and successful grassroots movements and to construct legal ways to linking right-to-the-city claims into the UN's human rights framework (UNESCO 2006).

While this new spate of interest is exciting, for the most part, the work has not systematically elaborated just what the right to the city entails, nor has it extensively evaluated the consequences the idea would have for empowering urban residents. While rising interest has produced greater detail through accretion, the right to the city is rarely engaged in depth (exceptions are Dikec 2001; Purcell 2002a; Fernandes 2006). To be clear, the work is innovative, stimulating, and welcome. However, as yet it falls short of a careful exposition and evaluation of the right-to-the-city idea. We lack a comprehensive explanation of what the right to the city is or how it would challenge, complement, or replace current rights. And we are left without a good sense of how the right to the city might address the specific democratic problems associated with urban neoliberalization. For example, a 2002 conference in Rome devoted to rights to the city produced many excellent papers, but virtually all of them failed to discuss the idea in detail (Rights to the City 2002). Almost none of the papers defined what it meant by the concept, and each seemed to take it as a given that more rights to the city were desirable. Other papers either failed to mention the concept altogether, or mentioned it only obliquely. That lack of engagement is quite common. A gulf exists between the frequency with which the right to the city is mentioned and the depth with which it is explored. In some ways, the "right to the city" has become something of a catchphrase; we have only begun to critically examine its potential for contributing to a renewed urban democracy. While uncritical evocations can be rhetorically and politically useful at times, my concern is that if we offer only a latent sense of what we mean by the right to the city, the concept will become steadily more amorphous and unhelpful, and it will fall into disuse without having been critically evaluated.

In an effort to begin a more thorough examination of the right to the city, the next section returns to Henri Lefebvre's writing on the subject. I use Lefebvre's ideas as a foundation to elaborate my interpretation of the right to the city. The interpretation that emerges in the following pages is based on a close reading of Lefebvre, but it is not meant to be simply a faithful reproduction of his argument. My goal is to draw heavily on Lefebvre to generate my own detailed interpretation of the idea. I think Lefebvre is a logical place to ground that project, since he is one of the original authors of the idea. Moreover, he offers both the spatial and urban conception of democratic politics that is lacking in so much democratic theory. Lastly, his work more generally offers a distinctly optimistic and creative way to imagine urban futures, precisely the kind of project this book advocates.

It is also important to be clear that in drawing on Lefebvre I do not mean to suggest we must establish and hew to an orthodox interpretation of the right to the city. We need to retain a multiplicity of interpretations. My dissatisfaction with existing usage of the term is not that there are many interpretations, but that they tend to be insufficiently explicit in their details. Our explorations of the right to the city need a vigorous debate about the specifics of the idea. That debate should not produce orthodoxy; rather it should produce a more robust set of possibilities for what the right to the city can mean. A central strength of the idea, I hope to show, is its ability to be open and flexible so that it can be mobilized in a variety of urban political contexts. My goal is to produce an idea that can serve as an organizing inspiration for a range of movements. But at the same time, as I have argued, there must be limits to that openness. We must begin to get specific about what the right to the city should entail. Otherwise we begin to lose the power of the idea. What I am trying to create, then, is an interpretation of the right to the city that offers much more detail than is typically offered, but that remains significantly open. It is a difficult trick, and one I may not entirely pull off. But it is one we must work toward if the right to the city is to provide a useful linchpin that can help bring together coalitions of democratic movements.

Lefebvre's Interpretation

As with most of his work, Lefebvre's portrayal of the right to the city is complex, fluid, and open to interpretation. It is fair to say that his articulation is designed more to evoke possibility than to unequivocally delineate his position. Lefebvre's right to the city is worked out most fully in *The Right to the City* (1996, 1968), but there is important supporting material in *Space and Politics* (1996, 1973), *The Urban Revolution* (2003), and *The Production of Space* (1991c).[4] I suggest that Lefebvre's right to the city is a claim that radically rethinks the social relations of capitalism, the spatial structure of the city, and the assumptions of liberal democracy. His right to the city is not a suggestion for piecemeal reform, although it does not necessarily exclude that as a tactic. Key to that radical nature is Lefebvre's explicit focus on urban space. The "city" he evokes is not merely material space, but a holistic sense of urban space as physical context, as social relation, and as everyday life. In his work in general and especially in *The Production of Space*, Lefebvre takes an extremely expansive view of space that includes what he calls perceived space, conceived space, and lived space (1991c). Perceived space refers to space as it is experienced in everyday life by its inhabitants, as when, for example, a person drives through an intersection every day and reaches the conclusion a four-way stop sign is needed. Conceived space refers to abstract and technical constructions of space, often associated with professionals and development firms, as when a traffic planner runs a flow analysis and concludes that same intersection needs a stop light. In addition to this binary opposition, Lefebvre posited a "third" space, a comprehensive lived space that has the potential to hold in tension and reimagine perceived and conceived space (Lefebvre 1991c, pp. 38–9; Soja 1996). Space is therefore bound up together with everyday life, with social relations, and with political struggle. It is socially produced, both through everyday life and through political struggle (Gottdiener 1994). Under the social relations of capitalism, conceived space, with its rational–technical reduction of space to a Cartesian grid, occupies a privileged position, one Lefebvre is concerned to undermine. Conceived space facilitates the marketization of space, the reduction of space to a measurable entity to be valued as property. Resistance to capitalist urbanism, for Lefebvre, requires a spatial resistance to challenge the hegemony of conceived space and to imagine more fully human alternatives. For him all political struggles are therefore always spatial ones as well. Producing and reproducing urban space, for Lefebvre, necessarily involves reproducing the social relations that are bound up in it. The production of urban space therefore entails much more than just planning and developing the material space of the city. It involves producing (and reproducing) all aspects of urban life. For Lefebvre (1996, p. 158), then, "the *right to the city* is like a cry and a demand" not just to a particular urban geography, but to "a transformed and renewed *right to urban life*." And urban life, for Lefebvre, is more than just life in a particular place. The city is a "social centrality," an epicenter and privileged confluence for social interaction and human creativity (1991c). The right to access that centrality is essential both for resistance movements and for each person's human flourishing.

Lefebvre's criticisms of the conceived space of professionals give rise to an interest in the perceived space of everyday life as a site of resistance. Here he focuses on the routines of inhabitance, of living one's life in the city (Lefebvre 1991a). His right to the city is designed to further the interests "of the whole society and firstly of all those who *inhabit*" (Lefebvre 1996, p. 158). Urban dwellers inhabit the city, professionals conceptualize it. In contrast to conceived space, which routinely ignores the complexities of daily inhabitance, the right to the city

> should modify, concretize and make more practical the rights of the citizen as an urban dweller (*citadin*) and user of multiple services. It would affirm, on the one hand, the right of users to make known their ideas on the space and time of their activities in the urban area; it would also cover the right to the use of the center, a privileged place, instead of being dispersed and stuck into ghettos (for workers, immigrants, the "marginal" and even for the "privileged").
>
> (Lefebvre 1991b, translated in Kofman and Lebas 1996, p. 34)

In contrast to the current situation in which planners, architects, developers, and other experts hold tremendous power over the production of urban space, Lefebvre imagines a central role for the *users* of urban space to determine its future. The daily routines of inhabitants shape urban space as an oeuvre, as a collective work of art. As they work at their jobs, share a break together on the sidewalk, shepherd their children to the park, look for a place to get out of the rain, wait for the bus, sit outside at the café, stand in line for the shelter to open, shop at the store, and move about the city, walking, riding, or driving, inhabitants are carrying out daily acts of survival. They are actively *inhabiting the city*. For them to be able to inhabit well— to realize a full and dignified life—the city must provide them what they need: employment, shelter, clothing, access to healthy food, and all manner of services, like child care, transportation, water, sewerage, education, open space, and the like. Claiming a right to the city is claiming a right to inhabit *well*, to have reasonable access to the things one needs to live a dignified life. For users, the city is a creative and collective human project, one that thrives on interaction, cooperation, and affective relations. For capital, on the other hand, the city is a strategic site for accumulation (Castells 1977). Urban space is a commodity to be owned, by holders with property rights. Its purpose is either to be valorized in its own right, or to serve as a platform on which accumulation can occur. The right to the city is thus a counter-claim against that neoliberal idea of urban space; it engages a debate over what the city is *for* (Castells 1977).

One component of the right to the city is what we might call a right to *appropriation*. In its most straightforward sense, appropriation can mean the right to be physically present in already-existing material space. Thus when the demonstrators in Seattle entered the no-protest zone the City of Seattle attempted to construct around the WTO convention, they were claiming a right to be bodily present in the city. While this may be the most basic facet of a right to appropriation, it is not at all unimportant. Otherwise, the police would not have gone to such extreme measures to quash the demonstrators' right to assemble, using tear gas, batons, and rubber bullets. Don Mitchell (2003) convincingly makes a similar argument, that it is critical to be able to be bodily present in urban public space. He sees it as a necessary platform on which counter-movements can demonstrate their strength and articulate their message. More urgently, for people without shelter public space is essential for survival. Anti-homeless, zero-tolerance policies that make it a crime to do things like wash, urinate, sleep, or even sit in public space are all typical of the disciplinary aspect of neoliberal urban policy. The right to appropriation is therefore also the right to use public space for survival. Less cruel and more subtle, but also important, is the way the privatization and commodification of urban space impinges on inhabitants' ability to carry out their daily routines.

Recently my wife and I took our twin two-year-old girls downtown to see the Christmas display at the big department store. It was cold, and as we waited for the bus home we tried to go inside to the common area of a nearby shopping mall to get warm. But the entrance to the mall was blanketed with signs making clear that the common area was for shoppers, not those wanting to wait for the bus in a warm place. The mall's "common area" was private property, and its owners allowed only customers to be present in its space.

But as important as the right to be present in material space is, Lefebvre's vision isn't only for a more user-centered design of concrete space. It is rather for a comprehensive reimagining of the city as a social and spatial entity. The right to appropriation can be conceived not just as the right to be physically present in existing urban space, but the right to a city that fully meets, above all other considerations, the needs of inhabitants. The employment, shelter, food, and other services mentioned above would be among the necessary elements of such a city. Appropriation in that larger sense would mean a right to a city where workers could make a short commute to work on frequent buses and come home to affordable, comfortable housing. It would allow child-care-givers to choose from several nearby parks within walking distance that offered kids spaces that stimulated their imagination. It would mean shoppers visiting a nearby grocery store that offered high-quality, reasonably priced food. It would mean a city without racial and class segregation that reinforced social inequalities. Certainly appropriation demands the right to be present in space, but it also requires the production of spaces that actively foster a dignified and meaningful life. It demands that the city, more than anything else, be *for* inhabitance.

Another element of Lefebvre's right to the city is what might be called a right to *participation*. When Lefebvre talks of participation he is referring to the ability of urban dwellers to participate fully in the many opportunities the city has to offer. He is concerned about the exclusion of inhabitants from those opportunities. A claim to participation therefore aims in part at a more inclusive city, one where opportunity is spread more equally across the population. That is the main way the United Nations project has mobilized the term, as a right to inclusive, sustainable, and socially just cities (UNESCO 2006). But of course participation also has a political element. It implies a sense of inclusion in decision-making, a meaningful say in all the processes that produce urban space. The right to the city is therefore not only the right to have the city produced to meet the needs of inhabitants, but also the right for inhabitants to participate fully in the production of urban space. It means, to use Dryzek's terms again, an extension of franchise, scope, and authenticity so that urban inhabitants are included meaningfully in decision-making about urban space. Certainly inhabitants, through their daily routines of living in the city, always play a central role in the production of the city as a collective product. But participation in this sense also claims the right to participate in the larger decisions that reshape the city.

The right to participation thus helps explicitly link the right to the city to democracy. The tradition of participatory democracy discussed in Chapter 2 bases its notion of democratic life on the meaningful political participation of citizens. The essence of democracy for participatory democrats is including citizens in the process of making political decisions. Participation both develops citizens' capacity for civic wisdom and produces wiser, more sustainable public decisions. For participatory democrats, however, those decisions are wiser because they are more fully in the common interest. Since the democratic attitudes I advocate above reject the common good as the ultimate aim of democratic politics, there is a need to go beyond this simple link between the right to the city and participatory democracy. We need to conceive of how networks of movements might claim a right to the city to resist neoliberalization and advance radical democratization. I develop that argument in more detail below. For now, however, it is enough to say that the right to the city demands that inhabitants play a central role in the decisions that produce urban space.

An Example: Brazil's City Statute

In order to better flesh out what the right to the city might look like in practice, it is useful to examine empirical cases. Examples of the right to the city in action are growing. Perhaps the example that best helps concretize the more abstract ideas above is Brazil's "City Statute" (the following account is taken from Fernandes 2006). In 2001, Brazil enacted Federal Law no. 10.257. It is designed to advance several principles associated with the right to the city: (1) the regularization of informal settlements (i.e. *favelas*), (2) the social function of urban land, and (3) the democratization of urban governance. The first principle articulates an initial step in mobilizing a right to the city. In rapidly urbanizing areas in the developing world, informal settlements are increasingly common and house a growing majority of urban inhabitants (Davis 2006). They are typically on the outskirts of cities, in the least hospitable environments. Because they are not constructed through sanctioned channels, they often lack basic services such as sanitation, sewers, electricity, etc. The City Statute calls for the "regularization" of these settlements so that inhabitants can be more fully included in the various protections and opportunities that the formal sector offers. Here the claim is a right to housing, but also to the many services that typically serve urban residential areas. Those claims for basic inclusion mirror the claims of homeless people in the industrialized world. They do not necessarily imply integration or absorption into the mainstream culture or economy. Rather they demand access to the basic material benefits that are afforded those housed and served in the formal sector. To some extent, because of the dire conditions in *favelas*, regularization is a meager claim. As Don Mitchell (2003, p. 232) says for the case of homeless in the United States, "the right to sit on a sidewalk or to sleep in a park," however fundamental, is nevertheless "a pretty narrow right indeed." In addition to the critical first step of regularization, it is necessary also to make a claim for a fuller *equalization* of service opportunities within the formal sector: not just housing but decent and affordable housing, not just jobs but good jobs, not just transportation but efficient and convenient transportation.

The second principle in the Brazilian law concerns what Fernandes calls the "social use value" of urban space, which is opposed to its economic value. By that he means a recognition that all urban space serves a complex social function such that many different groups rely on it and use it in different ways. In regulating the development and use of urban space, therefore, the Brazilian law intends to balance the interest of property owners with the social needs of the population more generally. Property rights are not trumps; they are not even primary. Rather they are contingent on whether the property provides an adequate contribution to social needs. Under the City Statute, therefore, property rights are subjected to the counter-claim of appropriation. The surplus value that development produces is not necessarily captured by the property owner but can be mobilized to meet collective needs. Fernandes (2006, p. 46) opposes that goal with the traditional situation in Brazilian cities: where urban land is "conceived almost exclusively as a commodity, the economic content of which is to be determined by the individual interests of owners." The new claims of the City Statute directly draw on Lefebvre's concept of counterbalancing the right to property with the right of inhabitants to fully benefit from urban life and its multiple opportunities and services. It injects a strongly collective, social, and public understanding of urban space as a counterbalance to the privatized view of neoliberals.

The third principle draws both on Lefebvre's notion of participation and on the innovative participatory budgeting process in some Brazilian cities. It proposes an expansion of civil society and the increasing democratization of city management and policymaking. As with participatory budgeting, the hope in this case is that broader popular participation will help achieve other goals, such as regularizing informal-sector settlements and defending rights to the social use value of urban land. In Brazil such democratization involves the dismantling of clientelist political systems and the inclusion of historically marginalized populations into formal processes. It explores new initiatives to build a culture of democratic participation, one that produces meaningful outcomes for participants that encourage them to continue their involvement. It also involves a decentralization of decision-making authority from the central state to local governments, and from municipal governments to neighborhood units. At the same time, however, it recognizes the need for national-scale policymaking and coordination: a National Conference of Cities charged with upholding the City Statute has decision-making power over national urban policy.

Elaborated and Flexible

The last lesson from the Brazilian case is an important one. Traditionally, rights are understood as formal privileges that are codified in law. I have proposed an alternative understanding, in which rights are seen as political claims that movements make on the wider society. Those two views are different, but they are not mutually exclusive. It is of course a matter of record that rights claims often become codified, as with the civil rights movement in the United States in the 1960s. It is also possible to imagine a relatively more top-down approach in which elites create legal codes that can help social movements to press their claims. The Brazilian case presents a marriage of these two aspects of rights. Fernandes is clear that urban social movements, especially in the *favelas*, were the impetus for the City Statute. However, he is keen to stress the value of that law for subsequent movements to press their claims. He argues that the right to the city must be not merely a political notion, not merely a claim made by movements; it must also be a *legal* right as well. Such formal codification can partly relieve movements of the need to continually re-win their rights. At the global scale, such a legal approach would involve something like the recent initiative for a World Charter on the Right to the City. The idea of that initiative is to have the UN adopt the charter and absorb it into the human rights framework, so that the right to the city could be more effectively claimed as a human right under international law. At the urban scale, the strategy has so far been to develop charters for cities. That strategy has produced both specific charters, such as the Montreal Charter for Rights and Responsibilities and the European Charter for Safeguarding Human Rights in the City, and general statements about what those charters might look like, such as the Aberdeen Agenda produced by the Commonwealth Local Government Forum.

I take seriously Fernandes' stress on codification. However, I want to be very cautious about codifying the right to the city in law. I do not accept Fernandes' (2006, p. 41) suggestion that legal codification is a *necessary* step toward the right to the city. However, it is critical not to reject legal strategies either. More generally, it is critical to remain open to working strategically with the state to promote a radical democratic agenda. Urban social movements are integral both to the democratic attitudes I present above and to the way I have conceived of political rights. They must play a central role in pursuing the right to the city. But their strategic choices will vary according to their particular opportunities. There may very well be situations in which legal codification is the right choice to make. There are often situations in which working with or even inside the state is either necessary or even desirable. The innovative initiatives that have emerged in Latin America, for example, have been nurtured and even sometimes initiated by sympathetic governments who captured the state apparatus in electoral victories. Such governments are increasingly common across the continent, and they open up enormous opportunities. In the case study I discuss later, a coalitional social movement to ensure a Superfund cleanup that benefits inhabitants has worked at times closely and always shrewdly with state agencies and has creatively exploited legal opportunities and protections to press their claims. For other movements in other circumstances, such strategies might be impossible or undesirable. The greatest danger is that legal codification will dissipate the energy of social movements, that it will make them seem no longer necessary. That outcome is a severe threat to democracy. The right to the city must always remain primarily a political *claim* made by mobilized groups. That said, I argue that the right to the city can take on a legal existence, as long as a law is understood as a strategic tool for pressing the claim, rather than as a more-perfect evolutionary end-state that renders social mobilization no longer necessary.

The right to the city as I have presented it is intended to be one vision among many possible ones. I present a set of specific ideas that are based on Lefebvre's work and on the way the concept is being mobilized in places like Brazil. However, the elaboration of the right to the city must be a collaborative project. Its content must emerge from a collective debate among many different activists and scholars. More importantly, that content must serve as a set of *attitudes* that works much like the democratic attitudes I discuss above. They are not *principles* in the sense they are binding *a priori* rules that movements must follow. Rather they are attitudes; they are usual positions that social movements generally adopt. But they are always flexible; they can be adapted to new situations and change to meet the challenges of wider political-economic shifts. They offer movements a set of worked-out default positions that can be mobilized and articulated quickly without needing to be continually reinvented. But they should not be so rigid as to force movements into a strategy or agenda that doesn't make strategic sense. Adopted as a set of attitudes, the right to the city I articulate above is a promising way to flesh out a particular agenda for the democratic attitudes in Chapter 2. Its particular power, I argue below, is that the right to inhabit the city can be an effective linchpin around which networks of equivalence can act in concert. While not everyone inhabits the city in an *identical* way, neoliberalization threatens the inhabitance of a wide range of groups in a way those groups can construct as *equivalent*. Before exploring the right to the city's potential as linchpin, however, I want to explore some potential pitfalls the idea can generate.

Cautions

Despite the potential of Lefebvre's idea, there are also important cautions that should always accompany any discussion of the right to the city. The first is that even if the right to the city can be constructed as flexible, as I urge above, it can also be construed reductively. There is always a danger that the idea of inhabitance will be squeezed to fit into a particular social or political identity, and it will thus be limited in its potential to draw together diverse coalitions. While that may seem a generic concern that could be applied to any political idea, there are specific reasons to worry about it in the context of the right to the city, reasons that are to be found in Lefebvre's own writing. At times, Lefebvre seems to imagine the political identity of inhabitants to be tightly tied to the class position of urban dwellers. He writes that the right to the city must be realized by a "social force" that brings about a "radical metamorphosis" in society (1996, p. 156). He sees the primary engine of that social force to be the working class. He stresses the importance of workers to the point of claiming that "*only the working class* can become the agent, the social carrier or support of this [social force]" (1996, p. 158). While here he is not quite reducing the category of inhabitant to that of working class, he is having difficulty imagining any other political agent that can carry the right to the city forward. He is sketching a movement for the right to the city that is so dependent on an imagination of class mobilization that it could easily make the fatal mistake of the Old Left tendency to reduce other political and social identities to their class identity. The danger of that preference for class is that the way inhabitance gets framed is a critical question in any politics of the right to the city. The great potential of the idea is that it can serve as a strategic linchpin through which different groups can act in concert. But that is also its greatest challenge. Bringing movements together while affirming their irreducible difference is an extremely difficult political task. Any intimation of a transcendent preference for a particular political agent, like the working class, threatens to make that difficult task nearly impossible.

Another concern has to do with scale. There is a danger that the right to the city can be understood as somehow necessarily linked to the urban scale, that the right to inhabitance is somehow best suited to an urban scale as opposed to larger ones (Purcell 2006). Lefebvre does not take such a position explicitly. Indeed, in other writings (and to a lesser extent in the *Right to the City* itself) his vision of the urban is expansive, not parochial (especially Lefebvre 1991c, 2003). He understands the urban as a complex social and spatial ensemble that transcends the physical city to incorporate networks of actors and systems involved in multiple relations at many scales. However, it is worth taking seriously that he didn't write about the right to the *urban* (l'urbaine) or the right to *space* (l'espace), but about the right to the *city* (la ville). His other, more expansive conceptions are primarily of "the urban" (l'urbaine). When he discusses the right to *the city*, therefore, he seems to be primarily talking about a right to inhabit the physical city, rather than a right to "the urban" more generally. Augmenting that danger is the fact that Lefebvre's work is characterized by a longstanding fascination with everyday life, with the concrete acts and understandings of people engaged in ordinary routines. When he writes about the right to the city, he celebrates these sorts of quotidian routines—the commonsense experiences of perceived space—as a counterpoint to the detached, rationalized understandings of owners and planners. He clearly links those embedded experiences to the rights associated with inhabitance. Talking about the right to the city in that way, as worthwhile and evocative as it is, runs the risk of arguing that the right to the city must be mobilized on a highly local scale, at which it is possible for inhabitants to develop those intimate understandings of their environment. The city, in this understanding, would be the very *largest* scale at which a right to the city might be mobilized. While Lefebvre's other writing on the urban pushes him to reject such scalar limitation, there are legitimate concerns about his exposition of the right to the city. Moreover, it is certainly possible that others pursuing the concept will interpret the right to the city in that more local way. It is only a very short step from privileging inhabitants to privileging *local* residents. That would mean that neighborhood groups (not citywide bodies) should decide how neighborhood space is produced, since they most fully inhabit that space every day. Or, on a larger scale, it could mean that citywide inhabitants (rather than decision-making bodies at larger scales) should control the production of space in their city.

That approach is a clear manifestation of "the local trap," an error that assumes that the local scale has inherent qualities (for a more complete exposition of this concept, see Brown and Purcell 2005; Purcell and Brown 2005; Purcell 2006). Scale is better understood as a strategy: the outcomes that result from scalar reorganization flow from the *agendas* of the actors empowered by the strategy. Localization can produce desirable or undesirable outcomes. So, movements that pursue only localization will not only run the risk of getting counterproductive results, but they will also miss scalar strategies that might be more effective for them. Resisting neoliberalization requires movements to pursue political initiatives at a range of scales, and to do so flexibly and strategically. Different temporal and spatial contexts will require different scalar strategies. A right to the city limited to the urban or even the neighborhood scale would undermine that necessary flexibility. It is worth reiterating that I am not saying Lefebvre himself, if we take his work as a whole, would have affirmed a locally trapped right to the city. In fact, the opposite is likely true. However, there is a clear danger that the right to the city idea can be mobilized in locally trapped ways.

An analogous danger concerns the urban not as a scale (urban vs. national vs. global) but as a settlement pattern (urban vs. rural). The right to the city seems to suggest something magical, something critically important, about the city as a thing one must have a right to. It seems to suggest that a right to the countryside is not similarly important. Throughout Lefebvre and other work on the right to the city, it is possible to discern a sense that cities are indeed something distinctively important, that they offer a higher level of opportunity and human achievement than other places. Access to cities and their opportunities thus becomes particularly important, more important than access to other places.[5] In this case the danger is more subtle than in the case of scale. Cities *are* different from rural areas; they operate as particularly strategic centers in the global political economy. In important ways, claiming a right to *cities* has particular importance for resisting neoliberalization, as I suggest above, since the city has served as a key launching pad and laboratory for neoliberal policy initiatives. Moreover, as Saskia Sassen (1994, 2000) has made clear, particular cities serve as key nodes in the networked global economy. They house concentrations of the many command-and-control functions the very large corporations depend on to coordinate their highly complex and far-flung operations. Such global cities are therefore extremely strategic sites for resistance (Purcell 2003a). Despite that very real strategic importance, however, the danger is that the right to the city can be interpreted to say that there is some *necessary* relation between the city and democratic resistance (Low 2004). My concern is that when we emphasize inhabiting the city, we will mistakenly assume that the *city* is the point here, rather than *inhabitance*. I argue that *inhabitance* should be the heart of the alternative that the right to the city offers us. The city is important, but mostly as a strategic frontier in the struggle against neoliberalization. The larger claim we should make is for the right to *inhabit space*. We can and should make a more particular claim to inhabit the *city*, but only as a strategic part of that larger claim. We should, in short, heed Neil Brenner's formulation of Lefebvre's vision. As cities take on increasing strategic importance in the global political economy, Brenner argues (2000, pp. 374–5)

> urban social movements have acquired an "importance and resonance … on a world scale" (Lefebvre 1978, pp. 161–2): they do not merely occur within urban space but strive to transform the socioterritorial organization of capitalism itself on multiple geographical scales. The "right to the city" … thereby expands into a broader "right to space" both within and beyond the urban scale (Lefebvre 1978 pp. 162, 317; 1979, p. 294). Even as processes of global capitalist restructuring radically reorganize the supraurban scalar hierarchies in which cities are embedded, cities remain strategic arenas for sociopolitical struggles which, in turn, have major ramifications for the supraurban geographies of capitalism (International Network for Urban Research and Action 1998).

I would merely augment Brenner's evocation of Lefebvre's "right to space" by reiterating we must claim a right to *inhabit* space. But Brenner's essential point remains: we must imagine the city expansively, with Lefebvre, not only as a discrete territory and physical environment but also as a node in multiple networks at a variety of scales. The right to the city then becomes not just a claim to the concrete space of the city as a scale and settlement pattern, but a strategic claim in a broader movement for the right to inhabit not just urban space, but space in general.

Another important concern involves the distinction between inhabit*ants* as political subjects and inhabit*ance* as an agenda. Movements of inhabitants may or may not claim the right to inhabit space. Or they may claim that right and also make other claims. They can resist neoliberalization, but they can just as well support it, or be reactionary, or exclusionary, or racist. As Laclau and Mouffe (1985) stress, inhabitants, like any other political subjects, can construct their collective interests in a variety of contingent ways. The degree to which they claim the right to the city will depend on a variety of factors. I do not want to make an argument about false consciousness here. I do not mean to imply that inhabitance is the "true" agenda of inhabitants, and any other agenda they might pursue is a distortion produced by the cultural dominance of bourgeois ideas. That kind of essentialism can very easily produce a vanguardist approach, which radical democratic movements must by definition reject. Instead, I am arguing that we should not read off anything about a movement's agenda from the political subjects who people it. A movement of inhabitants is not necessarily pursuing a right to the city. In order to claim a right to the city, inhabitants must decide to mobilize, at least in part, around an agenda that claims the right to inhabit space. Moreover, in order for the right to the city to effectively resist neoliberalization specifically, the right to inhabit space should be understood consciously as a direct alternative to the right to own and profit from space.[6] Space-as-inhabited should be seen in opposition to the commodified space of neoliberalism. Movements of inhabitants can instead construct very different agendas. They can, for example, pursue the right of "people like me" to inhabit my neighborhood, and see that right in opposition to the right of "people not like me" to inhabit my neighborhood. In that case, as we will see in Los Angeles, we have a very different kind of inhabitant movement, a reactionary one that may or may not resist neoliberalization (Purcell 2001b).

The distinction between inhabitants and inhabitance is important because it helps us avoid the trap of assuming inhabitants will necessarily pursue a progressive and anti-neoliberal agenda of inhabitance. It also helps us see more dangers in any legal strategy to codify the right to the city. If we understand the right to the city purely as the inalienable right of particular individuals (inhabitants), it can be mobilized in a variety of ways incompatible with the agenda of inhabitance. If instead we see the right to the city primarily as a claim around which social movements can pursue a city that nurtures inhabitance, the goal becomes not so much legal codification as active and energetic *fostering* of the movements that claim a right to the city. A companion to such fostering would of course be active *opposition* to the movements of inhabitants that are working at cross-purposes to the right to the city.

A last concern about Lefebvre's right to the city is that he developed his concept in 1968. Neoliberalism was not ascendant in the way it is today, although the cracks in the Keynesian compromise were certainly becoming apparent. He did not have the benefit of being able to articulate how the right to the city might engage with the assumptions and policies of our current era of neoliberalization.[7] Moreover, Lefebvre developed his idea without developing explicitly how it might be mobilized as a specifically *democratic* movement. Lefebvre had a sense that "urban democracy" was linked closely with claims to the right to the city (2003, pp. 136–7, 142). He intimates (1996, pp. 75–6) that these are cut from the same cloth. But his focus is on other theoretical issues; he does not explore democracy as an object of analysis. He adopts the same kind of unexamined assumptions about democracy as many revolutionary democrats (discussed in Chapter 2): his democracy is taken to mean direct, participatory, and total democracy. It empowers the mass of people in all spheres of decision-making and therefore makes impossible the hierarchical and concentrated corporate power that characterizes contemporary capitalism. The absence of a sustained examination of neoliberalism and democracy in Lefebvre, neither of which constitutes any failing as a scholar on his part, should be taken not as a shortcoming but as an opportunity. They urge us to begin the project of weaving a concept of the right to the city into both an analysis of contemporary neoliberalization and a more explicit and critical analysis of democracy. My hope is that by taking seriously how both democracy and the right to the city meet the challenges posed by contemporary neoliberalization, we can more effectively mobilize their power and their promise.

DEMOCRATIZING CITIES

To begin that project, let me return to our starting point, the democratic attitudes that lack a spatial and urban element. Those attitudes embrace a specific vision of democracy, one not limited to the actually existing liberal democracy we currently experience. They reject the notion that all democratic politics must aim at the common good. Rather they embrace an agonistic model in which adversaries with unavoidably divergent interests struggle with each other to win a temporary hegemony that favors their agenda. It is a social-movement vision of democracy, one that imagines distinct movements that act together in networks of equivalence. Their methods mobilize (among other strategies) the discourse of rights, not as inalienable trumps held by individuals, but as political claims that movements make on society in the course of their struggle. Their goal is to undermine the current political hegemony of neoliberalism and establish a different one. They hope that new hegemony will involve radical democratization and radical equalization. In this vision, democratic relations are not limited to the formal political sphere but extend to all facets of social life, including the workplace and the family. Those more extensive democratic relations are simultaneously deepened by a radical social equality that at last makes meaningful the democratic promise of political equality.

The right to the city adds value to these democratic attitudes in at least two ways. First, as I say above, it offers a deeply spatial and specifically urban way to imagine the agenda that democratic movements can pursue. As they claim the right to inhabit space, movements struggle over the question of what the city is *for*. It is an age-old question, right at the center of Aristotle's *Politics*. All hegemonies must answer the question; they must install their answer as the commonsense answer. Successful counter-hegemonic movements must destabilize the dominant answer and articulate their own. Neoliberalism, for example, constructs the city as a strategic node in a network designed to maximize capital accumulation. Urban space is imagined to be owned property and its role is to contribute to economic productivity. The right to the city destabilizes that neoliberal answer and offers a distinctly new vision for what the city is for. The right to the city demands that we see the city first and foremost as *inhabited*. The city is a collective creative work by and for inhabitants. It depends on them, for they are its creative producers. But they also depend on the city: it is their habitat, the space of their everyday survival. Nurturing the relationship between city and inhabitant, therefore, and between urban space and inhabitance, becomes the principal imperative of urban politics. That new imperative supplants the neoliberal imperative of economic growth. But it does so through agonistic struggle: it supersedes the neoliberal vision, but it cannot expunge it.

Second, the right to the city adds value because it offers one way to construct a shared vision with which networks of equivalence can move forward. That vision rejects the neoliberal city-as-property and actively advocates the city-as-inhabited. The promise is that a range of urban movements can share an equivalent connection to the idea of the city-as-inhabited. That shared vision is vital for the most critical project that such networks must achieve: hold together while remaining distinct (Young 1999b). Recall that Laclau and Mouffe define a relationship of equivalence as two things that are simultaneously the same and different. They demand that we do not take a reductive approach, that each group, with their movement and agenda, must retain its autonomy, distinct perspective, and particular relationship to the polity. However, to avoid postmodern fragmentation that consigns each movement to relative powerlessness, some sort of coordinated action must be organized. That coordination is stressed by scholars like Piven and Cloward (2000) and Harvey (1996), although it is fair to say they privilege class as the basis of coordination. Instead, what is required is the development of *equivalence* that does not reduce movements to their class (or any other) positions. Movements must not only confront the long-standing questions of how to construct shared values and build group solidarity (Miller 2000; Jenkins 1983; Calhoun 1988), they must now do so across broad networks of diverse movements, each themselves internally differentiated, without reducing that difference. Moreover, all the other challenges that confront social movements remain, like how to mobilize resources like money, leadership, and networks (Gamson 1990; McCarthy and Zald 1977) or how to seek and exploit political opportunities (McAdam 1996; Tarrow 1996; Kriesi 1995).

Achieving the difficult trick of equivalence requires that movements construct a basis for it, a narrative that allows them to see themselves as holding an equivalent grievance under the current hegemony, and an equivalent hope for an alternative future. Negatively defined, equivalence might entail a shared opposition to the neoliberal city-as-property. Groups against environmental injustice in non-white communities, or the destruction of wildlife habitat, or gentrification and dispossession in poor neighborhoods can construct equivalent (though different) objections to the neoliberal insistence that economic development is the city's predominant purpose. That kind of equivalence is being developed in the case of the Duwamish River cleanup, which I examine in Chapter 4. Opposition to the neoliberal agenda, therefore, can be one way to construct equivalence. But equivalence can and should also involve a positive element, a shared alternative to the current hegemony. That is where I think the right to the city can be most useful. It offers real promise as a starting point for constructing equivalence. The urban social movements I mention above can all potentially decide that they share an equivalent interest not only in opposing the city-as-property, but also in *advocating* for the city-as-inhabited, for claiming the right to inhabit urban space. Communities concerned to defend themselves from environmental injustice can construct their neighborhood as inhabited, by people whose health and well-being should be paramount. Groups wishing to restore the health of a polluted river can insist that the river is inhabited; they can emphasize the needs of the (generally non-human) species that rely on it. Groups opposing gentrification and displacement can emphasize the everyday needs and routines of low-income inhabitants who often have few other options for affordable housing in the city. And one can imagine how this model might work for other movements, such as those concerned with community development, or affordable housing, or public health, or the needs of homeless people. Each of these groups and movements has a different relationship to urban space, they inhabit it in different ways, and they face different challenges. But they share an equivalent concern for preserving the right to inhabit, as well as an equivalent grievance against the right to property and accumulation.

Recently, for example, some initial steps were taken to explore how inhabitance might be equivalence. In Los Angeles in January 2007, over thirty community-based organizations came together to discuss the question of whether the right to the city might be an effective "frame," as they put it, for seeing the overlaps among their diverse agendas. The organizations were involved in struggles over displacement, women's rights, worker's rights, gay and lesbian rights, cultural preservation, racial segregation, and the like. But there was a hope among many that the right to the city frame could link women, people of color, gay men and lesbians, and workers through common urban struggles, struggles *against* urban space as property and *for* inhabitance (Samara 2007). While the conference was merely an initial step, it revealed promise. It remains an open question whether the right to the city can be an effective basis for equivalence. And it is probably not going too far to say that that question should be the central one as we evaluate the right to the city. Can it be a frame for constructing equivalence in the way the Los Angeles conference participants hoped? The fact that inhabitance opposes neoliberalization in fundamental ways is important, but if the right to the city cannot serve effectively as a frame for networks of equivalence, then its promise is greatly diminished. While the conference participants used the word frame to discuss how movements can construct equivalence, I have been using the word linchpin. In doing so, my aim is to use metaphor to better explicate the structure of the imagined democratic movements. I am not using words like hub or pivot or fulcrum in order to steer clear of the connotation of centrality they carry. I am rejecting the image of movements all hooking into an identical center in order to join together. Rather, I want to evoke the image of a decentered (yet coordinated) network, where each movement joins directly to all the others.[8] Each group, in concert with the others, constructs a notion of equivalence, and that notion is a linchpin each uses to secure itself to all the others. There are thus many linchpins dispersed throughout the network, not a single hub at the center. Each linchpin is equivalent to the others. That is, they are the same idea that is constructed differently by each group. Each group's linchpin is cut from the same material, but it is shaped by the group to fit their particular linkage to the network. Also, each linchpin can be removed by its group, and the group can exit without the rest of the network falling apart.

An elaborated but flexible notion of the right to the city has the potential to serve as a common starting point from which groups construct an equivalence through which they act in concert. Groups would not simply *adopt* the right to the city as their shared agenda. Rather they would each see something both relevant and inspiring in the concept, enough to commit to building a shared vision together with other groups. To reiterate Laclau and Mouffe's argument, that shared vision is not an expression or a discovery of a deeper political essence that each group shares. Rather the vision is the result of conscious action by particular groups to construct it as equivalent (Laclau and Mouffe 1985, p. 65). It is not, therefore, the result of political archeology to uncover what is already there. It is the result of political struggle to forge a strategic linchpin that resonates equivalently (but not identically) with the challenges facing each group. So the right to the city cannot be a fully formed agenda, it can only be an initiator, a catalyst, for a process of political construction.

Just what that process might look like then becomes an empirical question. It must be worked out by political movements in specific contexts. But the utility of these theoretical explorations does not end with the beginning of that political process. An ongoing dialogue between action and ideas is both possible and necessary. That holds true not just for radical democratic politics and the right to the city, but also for the other democratic traditions and their relationship to neoliberalization. The relationship between democratic attitudes and neoliberalization can only be fully grasped on the ground, in empirical context. Therefore, the next chapter is an investigation of a spectrum of different urban-political case studies from Seattle and Los Angeles, each of which sheds a different light on the relationship between urban democracy and neoliberalization.

CHAPTER 4

ON THE GROUND IN SEATTLE AND LOS ANGELES

The goal of these case studies is to better conceptualize the complex relationship between democracy and neoliberalization. I am exploring empirical cases in order to speak to my theoretical argument—to sharpen it and explore its limits. Beyond the book, however, the relationship is not merely one-way: theory has an important illuminating role as well, to offer insight and suggest new ways of looking at the world that movements can benefit from. Of course, in the bigger picture the relationship is a constantly developing dialogue, in which theory helps shape practice, and practice shapes theory. While that is not a new insight, it is all too frequently forgotten, both by theorists who do not sufficiently ground their ideas in concrete politics, and by activists who see abstract theory as irrelevant to their cause.

The chapter includes several cases. Each is intended to illuminate a different aspect of the relationship between democracy and neoliberalization. The first, concerning the South Lake Union neighborhood of Seattle, is a case that follows quite closely the dominant narrative in the political economy literature: pressures of competitiveness mean democratic decision-making is suppressed in favor of more expedited procedures. The outcomes strongly favor the interests of capital. The second case, about the revisioning of Seattle's waterfront area, illustrates some of the limitations of a deliberative-democratic response to neoliberalization. The third case is about the politics surrounding a Superfund cleanup on Seattle's largest river. My account of this case is a bit more detailed because it helps us understand better how radical democracy and the right to the city might operate in urban politics. It is a hopeful story, one that suggests great potential for networks of equivalence, even if that potential is only beginning to emerge. The fourth case examines homeowners' movements in Los Angeles and reveals the complexity that characterizes movements of inhabitants. They are not necessarily progressive, and they do not always pursue an agenda for inhabitance.

SEATTLE IN BRIEF

Before examining each of the three Seattle case studies, it is worth saying a bit about Seattle's political-economic context more generally. In many ways, Seattle can be considered a global city. It is not enormous like New York, London, and Tokyo; the Metropolitan Statistical Area has only about 3.2 million people, making it the 15th largest in the United States. But it is home to a number of economic enterprises that are central to the emerging global economy. Moreover, Seattle is the headquarters of a number of huge multinational corporations, like Microsoft, Boeing, and Costco, all of which are among the largest 140 corporations in the world, grossing $40 billion, $55 billion, and $53 billion respectively in 2005 (Fortune 2006). Microsoft, of course, is the most visible example. It is a central player in the global information economy. A number of other major firms, like Amazon.com and Adobe Systems, are also prominent in the local dot-com economy and in the global information economy. Other firms, like Costco and Starbucks, are central players in the global retail economy. Another Seattle firm, Washington Mutual, is by far the biggest savings institution in the United States. The area also has a historically and still very large primary extractive economy, mostly in softwood timber. Weyerhaeuser, one of the largest such firms in the world, anchors that sector. Seattle also has key players in aerospace. Although Boeing recently shifted their headquarters to Chicago, the region still hosts an enormous aircraft manufacturing industry. Each of these sectors—high tech, retail, timber, and aerospace—has numerous ancillary firms that are linked into local economic networks. Moreover, all of the very large anchor firms, except Boeing, maintain their headquarters in the Seattle metropolitan region. In short, for a city its size, it has an extremely productive economy and is a prominent node in many different global economic networks. However, despite its productivity, the local economy is also marked by fairly strong boom-and-bust tendencies. Aerospace in particular has always had noticeable and sometimes painful swings. The recent dot-com bust had acute local impacts. Boeing's 2001 emigration, therefore, as well as their recent flirtation with other sites as possible places to build their new 787 line (they eventually chose nearby Everett), has created a strong sense of urgency to keep and attract capital, despite Seattle's wealth relative to other U.S. cities its size.

In the midst of that relatively robust economy, the area in general and the City of Seattle in particular have developed a political culture that prides itself on democratic process. The culture is strong enough that it has produced a sobriquet: "the Seattle Way." In its positive connotation, the Seattle Way values popular participation, transparent process, and meaningful debate. More negatively, it has been decried as a culture that values process and debate over results, that bogs down and can't get important things done (Van Dyk 2005). One example that is both a root and product of the Seattle Way is an innovative approach to neighborhood planning, begun in 1995, that gave significant power to citizens' councils to devise their own plan for their neighborhood (Ceraso 1999). Local residents went through a process of learning about planning process, facilitated by experts, then deliberated toward a shared neighborhood plan. They then proposed the plan to the City Council to be adopted. Each of the 35 neighborhoods produced a plan, and the City adopted them all. Partly as a result, the Seattle Way places significant value on grassroots, bottom-up decision-making (Diers 2004).

Also central to the Seattle Way is a deliberative, cooperative approach to politics that places great importance on the common good, on consensus, and on civil dialogue among citizens. That deliberative political culture is embedded in the wider culture of Seattle's culturally dominant white middle-class: reserve, politeness, deference, and good manners are esteemed while boldness, confrontation, and aggression are reproved. The dominant population in Seattle defines itself culturally in opposition to the frank and in-your-face jostle of cities like New York. Seattle is a big city, but its mainstream culture is much more that of a small Midwestern city like Omaha or Iowa City. As a result, its political values stress cooperation, consensus, reserve, and civility. Confrontation, struggle, and conflict are avoided. As a result, the political culture esteems deliberation as the highest form of democracy. While most of the decision-making structures remain, as in most cities, characteristic of liberal democracy (elections, representatives, checks and balances, etc.), deliberative democracy is seen by many as the way forward, the political cutting edge that can democratize the imperfect liberal system. Innovative new democratic initiatives in Seattle are almost all cut from the deliberative cloth. In that context, social mobilization among radical democratic movements is not only outside the traditional liberal-democratic mainstream, it is also outside the predominant deliberative-democratic alternative. Its politics are too self-interested, too confrontational, too, well, *loud* for the Seattle Way. It is in that context that each of the Seattle case studies sits. They are based on an ongoing long-term research project that has spanned the last four years. Like the Los Angeles study (below), the project has gathered data through observation, interview, and archival methods and analyzed them with grounded-theory techniques. All information whose source is not cited is taken from field notes from that project.

South Lake Union

> We are moving as fast as we possibly can. Holding onto land is not desirable when we can turn it into income—producing property as quickly as the market will let us do it.
>
> Ada Healey, vice president of real estate for Vulcan, Inc.
> (quoted in Mulady 2004b)

The South Lake Union (SLU) area of Seattle is currently undergoing a significant redevelopment that follows quite closely the narrative of the political economy literature on neoliberalization. SLU has long been a fairly low-density mixture of warehouses, car dealerships, marine-commercial, light manufacturing, and, in its Cascade sub-section, residents with relatively low incomes (see Figure 4.1 and Figure 4.2). It is located just a bit north of downtown, in a relatively central location. Just to its south, between SLU and downtown Seattle, the Belltown neighborhood recently underwent a thoroughgoing gentrification and is stuffed with expensive condominiums, high-end restaurants and retail, and trendy nightspots. Up the hill to the east from SLU is Capitol Hill, one of the city's highest-value residential districts. To SLU's northwest is Queen Anne, also mostly high-value residential. To its west is the Seattle Center, Seattle's designated hub for civic events, and home to the Space Needle, Paul Allen's Music Museum, and the city's NBA arena. In short, seen through the lens of highest and best use, and considering its surroundings, SLU is currently quite a lot less valuable than it could be. It is a relative exchange-value desert (see Figure 4.3) in the middle of a long-booming real estate market whose values have increased steadily at an average annual rate of 7.3 percent over the last 12 years (Puget Sound Regional Council 2005).

Partly driven by these low values, the city's largest real estate firm, Vulcan Inc., which is owned by Microsoft billionaire Paul Allen, has begun a very large-scale project of speculative investment in SLU. Over the past several years it has come to own a significant portion of the property in the area (see Figure 4.4). Vulcan's project is multifaceted, but one central element of its vision is that the area will be redeveloped into a global "technopole" for biotechnology (Scott 1993). Such biotechnology hubs are an increasingly common neoliberal growth strategy (Larner 2005). In the SLU case, the technopole would be supported by a number of biotechnology firms already operating in the area, as well as the University of Washington (about a mile away in the University District), which has a particular expertise in biotechnology research. Vulcan is not merely a real-estate speculator; it is also a land development firm. Their goal is not

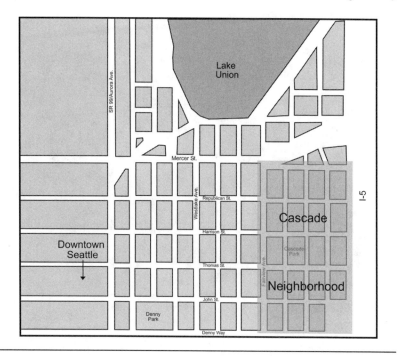

Figure 4.1 South Lake Union, showing Cascade neighborhood

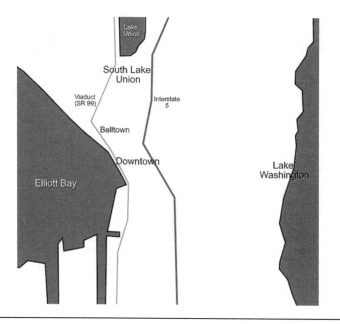

Figure 4.2 South Lake Union in context

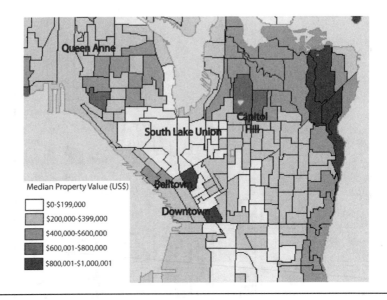

Figure 4.3 South Lake Union in the context of surrounding property values

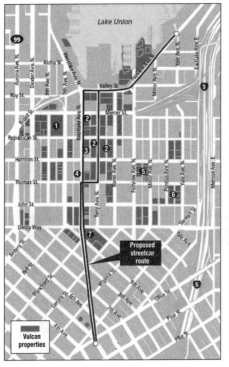

The Seattle City Council will consider more than a half-dozen proposals related to South Lake Union to make it easier to develop the neighborhood into a biotechnology hub. Billionaire Paul Allen's Vulcan Company is leading the posh. Allen owns 58 acres in South Lake Union, and Vulcan plans to build more than 10 million square feet of residential, office and commercial space.

CURRENT VULCAN DEVELOPMENTS
This yea, companies have moved 550 employees into South Lake Union buildings owned by Vulcan.

1 815 Mercer St.: University of Washington plans to move about 300 employees into 105,000 square feet in the renovated Washington Natural Gas "Blue Flame" building in December. The building is leased by the university.

2 401 Terry Ave: Rosetta Inpharmatics-Merck moved about 300 employees into part of the Interurban Exchange complex owned by Vulcan.

3 428 Westlake Ave. N.: Proposed office building, to be occupied in part by CollinsWoerman Architecture.

4 307 Westlake Ave. N.: Seattle Biomedical Research Institute moved in 170 employees; Children's Hospital brought 60 workers.

5 301 Minor Ave. N: The 162-unit Alcyone apartment complex opened in May.

6 223 Yale Ave. N.: Headquarters for NBBI and Skanska. The development that will include 180 residential units, retail and office space.

7 2200 Westlake Ave. N.: Commercial and residential complex anchored by Whole Foods and the Pan Pacific Hotel.

Figure 4.4 Vulcan holdings in South Lake Union (Source: Mulady 2004b)

Buildings planned, under construction or recently built

Figure 4.5 South Lake Union, showing completed and planned development (Source: City of Seattle, Department of Planning and Development)

merely to sell their holdings at a higher price, but, in concert with the city and some other land development firms, to develop the area into a range of uses, including office, commercial, residential, and parks (see Figure 4.5). They have already developed several properties in the area, most according to a distinct new-urbanist ideal: mixed-use, green/sustainable building technologies, and dense, walkable urban form. They see their SLU project as nothing short of a grand enterprise to reimagine the city. "Rethink urban," their website urges us, "that's what Vulcan is doing at South Lake Union" (see http://www.vulcanrealestate.com).

Not surprisingly, the comprehensive redevelopment of a centrally located, relatively depressed area of the city is very appealing to city and state government. The technopole vision is the expressed goal of both Seattle's mayor, Greg Nickels, and Washington State's governor, Christine Gregoire. While some members of the City Council have been a bit more skeptical, generally government has been an enthusiastic backer of the redevelopment vision. There has been a strong convergence of agendas between the private sector and leading figures in municipal and state government. The degree of cooperation is so high that some have intimated there is corruption at work. Some current Vulcan employees are former staffers for the mayor and city council, suggesting Vulcan's desire to maintain easy access to and strong relationships with decision-makers (Mulady 2004a). While Vulcan is certainly working to secure influence at City Hall, nevertheless it would be missing the point to conclude that undue influence is behind the City–Vulcan partnership. Rather, the agendas of the City and Vulcan are converging around the desire to create a biotechnology boom, densification, and a steep rise in property values in SLU. Ada Healey, vice president of real estate for Vulcan, argued that the redevelopment of South Lake Union "means more jobs and tax revenue for the city" (Mulady 2004b). City officials, especially the mayor's office, echo that argument eagerly. Increased property values in South Lake Union, the mayor believes, would result in up to $133 million in new tax revenues over the course of a 20-year redevelopment (Ceis 2004). He also makes regular claims about the job growth in SLU. A recent press release says that 499 new jobs have been created to date (Office of the Mayor 2004a) and claims that as many as 76,000 more are on the way. "Jobs in South Lake Union," the mayor claims, "are good for Seattle, good for the region and good for the entire state" (Office of the Mayor 2004b). While the City Council has been a less aggressive booster in SLU, in 2003 they nevertheless unanimously passed a formal resolution stating their "commitment to making the South Lake Union area the region's most competitive location for biotech research and manufacturing, clean energy … other high-tech research and manufacturing, and other innovative entrepreneurial high-tech industries" (City Council 2003). Those converging agendas are the main reason the City and Vulcan have so clearly coordinated their efforts. In essence what has emerged is an informal but nevertheless strong public–private partnership, a common neoliberal governance structure to promote an agenda of economic growth.

To begin the redevelopment, Vulcan bought many of their current holdings from the City at a bargain rate. The City then rezoned that property to make it more amenable to Vulcan's vision. The changes included office space for biotechnology, rezoning for mixed use developments, and increased density.

Moreover, the infrastructure in the area is generally quite dilapidated, especially the roads, and so the City (together with other government agencies) has pledged to carry out significant improvement projects. Such improvements will of course require massive public investment. Estimates vary as to the amount of this investment. The City suggests it is around $420,000,000, while at least one activist counters it is closer to $1 billion (Mulady 2004b). As we saw, the City believes that any investment in infrastructure will be repaid many times over in terms of jobs and tax revenue if SLU becomes a global leader in biotechnology. I should point out that no matter what the development plan, such infrastructure improvements are badly needed in SLU. Spending public money to improve SLU roads, for example, is perfectly reasonable, given their manifest disrepair. However, the needs have existed for a long time; it is clear that the City is committing funds for improvement *now* because it wishes to foster comprehensive redevelopment and a biotech hub. For the same reason, it is spending funds *here* rather than in other parts of the city with similar needs, again because its spending is driven by the desire to foster economic development, not to meet the needs of inhabitants.

Mostly as a result of the partnership's efforts, a clear narrative has emerged to advocate for the redevelopment of SLU. The narrative begins with the story of the fragile regional economy. Boeing, the region's aerospace anchor, is carrying out an ongoing disinvestment, moving both headquarters and some production, that will deeply impact not only Boeing employees, but also the industry's many ancillary businesses. Adding to the sense of impending doom is a skittishness born of the recent dot-com bust of the early 2000s. To an extent Microsoft and the high-tech economy eases foreboding about aerospace disinvestment, but the dot-com bust revealed that much of that economy is fragile as well, vulnerable not so much to firms leaving but to firms collapsing. As a result of that instability in existing economic sectors, the narrative goes, we need to hit on a new fix, a "next big thing," that will ensure continued economic growth. In Seattle (and elsewhere), many see biotechnology as the most likely next big thing, one the state should invest significant public money to stimulate. One plan suggested spending $250 million in state money, which a state senator said was needed "to grow our economy." "The evidence is there," said a member of the state house of representatives, "on the return that we get with that type of investment" (Dietrich 2003). Capturing a leading center of biotech can save Seattle from the vagaries of the other sectors. Feeding this narrative, Mayor Nickels offered one of his many promises about job creation by promising that biotech will be "equivalent to another Microsoft" (Office of the Mayor 2003). "The revitalization of this neighborhood," he added later, "represents an opportunity for Seattle as significant as the day Bill Boeing decided to build airplanes" (Tindall 2004). Those are enormous promises. They are also irresponsible, since it is quite difficult to know if biotech in

SLU will become anywhere near as significant (regionally or globally) as Microsoft or Boeing. There is in fact much debate about the multiplier effects of biotechnopoles (Genet 1997; Autant-Bernard *et al.* 2006). But "another Microsoft" or Boeing are references that resonate powerfully in the local imagination.

Buttressing that narrative about the need for economic strategies to insulate the regional economy is another narrative about SLU itself. The second narrative characterizes South Lake Union as dead: a "moribund warehouse area," in need, quite literally, of revitalization, of an injection of life (Rivera 2004). Vulcan reiterates the narrative of revitalization insistently. Redeveloping the area, they claim, will bring "revitalization," "a livelier 24/7 urban lifestyle," "a more vibrant … atmosphere" that has already "energized the neighborhood" and produced an increasingly "thriving" and "lively" community (Healey 2006, who is particularly relentless on this point). The patient was dying, they insist, and redevelopment is helping to bring it back to life. South Lake Union's woes are characterized in part as economic. It is an exchange-value vacuum amid prosperity. Seen through the lens of highest and best use, it is a virtual wasteland, an "overlooked" (Tindall 2004) and "underutilized urban area" in need of an infusion of economic value (Healey 2006). But the area is also framed as dead in another sense, as empty of actual living people. We must transform it into a place "where people can live, work and play" (Fred Hutchinson Cancer Research Center 2004). Of course it already *is* a place where people can and do live, work, and play. But such narratives of revitalization, of future vigor, work to erase current SLU inhabitants from the imagined landscape. To the extent they are acknowledged at all, they are part of the dying past that desperately needs rejuvenation. According to the 2000 census, there are about 1,500 people living in the area. Those who have not arrived with the recent beginnings of gentrification tend to have low-to-moderate incomes, living in rent-supported apartments, co-ops, and other kinds of low-rent arrangements. The area also houses a dense cluster of social services, including alcohol and drug treatment, foster care, refugee resettlement, mental health counseling, and basic services for homeless people. The old-line residents and the social services are organized into a neighborhood group, the Cascade Neighborhood Council. Their politics are strikingly left-wing for a neighborhood group. They are committed, for example, to ideas like human rights for homeless people in the area, to sustainable practices in the neighborhood community garden, and winning legislative recognition for gay-marriage. They are of course concerned about the ongoing gentrification of the area, but like most such groups they are also concerned about everyday questions such as getting people to come out for a neighborhood carnival organized by the People's Center, or what to do about an increase in dog feces on neighborhood sidewalks. Those residents and social services, their everyday routines and concerns, belie the narrative of a "moribund" SLU. Once such people and their lives are imagined away, however, redeveloping SLU appears to cause minimal disruption, while promising tremendous economic vitality. Vulcan insists that "new pioneers" are beginning to people South Lake Union, which is seen as "the new urban frontier," just as Neil Smith (1996) has argued. Vulcan believes that those pioneers, along with "visionary leaders," are beginning to create a new South Lake Union that "will help revive and drive Seattle's economic engine for the next century" (Vulcan Real Estate 2007). And that economic engine (a frequently heard phrase) is absolutely necessary, they claim, because our regional economy must continually grow to remain globally competitive. Taken together, those narratives lead us inexorably to the conclusion that we must, for the good of the region as a whole, support a comprehensive redevelopment of SLU to create a biotechnology hub and regional economic boom.

Those local narratives are supported by the wider neoliberal narrative of global competitiveness. Recall that the city council's resolution explicitly called for SLU to become a "competitive location for biotech … research." Later, when the council debated whether to relax height restrictions in SLU to accommodate the needs of biotech offices, an executive at Seattle Biomedical Research Institute argued that the changes were "crucial to keeping Seattle competitive with other regions." Another from the Fred Hutchinson Cancer Research Center, the most prominent biotech firm in SLU, agreed: "these changes will make the area more competitive and will attract biotech jobs" (Young 2003). Because biotechnology is the hot new industry, many cities around the world will compete to attract biotechnology investment. If Seattle wants to create a globally competitive hub, it can't be indecisive. It has to be assertive. "If we continue to be passive," an industry lobbyist urged "we'll be passed by the wayside" (Dietrich 2003). If it wants the economy to remain vibrant, the City can't afford the luxury of the Seattle Way. It can't afford to vigorously debate the various conflicts that such a massive redevelopment will inevitably produce. Not only do we need to redevelop SLU, we must do so quickly, with minimal dithering, and in a way that is most attractive to biotechnology capital. In the debates about easing height restrictions, advocates expressed a clear need to move forward quickly (Young 2003). One activist, speaking more generally about the SLU decision-making process, lamented the breakneck pace of decision-making, "Cascade and South Lake Union went to great lengths to develop their neighborhood plans. These plans are being disregarded by the mayor's fast-track efforts to develop South Lake Union" (Murphy 2003). The narrative of competitiveness forecloses democratic debate in at least two ways. The first foreclosure is carried out by the pincer movement of the narratives sketched above. They back every decision into a corner: we must redevelop, and SLU is the obvious place. What other choice do we have? A decision without any real options is not properly democratic. Neoliberal common sense thus squeezes democracy in precisely the way the political economy literature suggests. It is a testament to the dominance of the neoliberal narrative frame that even in a city like Seattle, with a large, globally connected, and fairly diversified regional economy, a narrative about the necessity of ceaseless economic growth can be so quickly mobilized and so easily become common sense.

Second, partly as a result of the perceived need to act swiftly, decision-makers have worked to keep the decision-making process in SLU quite closed. They have used existing liberal-democratic channels to expedite decisions, relying on, for example, council votes, bureaucratic offices, or the mayor's executive authority. For example, in order to accommodate growth, in 2004 the City Council redesignated the area as an "urban center" under state growth management law (Department of Urban Planning and Development 2005). That change allowed more density and secured better access to state transportation funding, which will help pay for the very expensive transportation improvements the area will need. That decision moved through conventional liberal-democratic structures, such as council committees, public hearings, and a full-council vote. Where public input has been allowed in the process, it has not used the innovative new processes of the Seattle Way. Instead it has involved more traditional, and only nominally democratic, forums such as the typical public hearing where individuals travel to City Hall to offer their opinion into a microphone, and decision-makers largely ignore it.

The input of neighborhood groups has also been carefully managed. The Cascade Neighborhood Council (CNC) has been effectively marginalized in favor of another group, South Lake Union Friends and Neighbors (SLUFAN). While CNC represents mostly low-income residents and social services in danger of being displaced by gentrification, SLUFAN represents primarily business and property owners in the area. Their board includes representatives from Vulcan, PEMCO insurance, the Seattle Times, and Fred Hutchinson Cancer Center. CNC does hold one seat on the thirteen-member board, but their lone voice is far outweighed by the many other, more powerful interests on the board. Currently, the board does not have a member who is a Cascade resident. Despite being overwhelmingly made up of business and property owners, SLUFAN has become the acknowledged representative of the "SLU community" in deliberations about the area's future. Not surprisingly, SLUFAN is very sanguine about the impeding redevelopment. CNC of course has a far more critical view of the dangers of redevelopment, but for the most part they have been shouldered aside.

That strategic redefinition of "the community" was particularly important because of SLU's redesignation to urban center. Because the new urban center designation is governed by state law, it required a revision of the existing neighborhood plan for SLU in order to bring it into phase with state growth-management requirements. In general, state growth-management law is concerned to prevent sprawl on the urban fringes. It is therefore eager to make denser the already-built central areas of the city. Urban centers are the places designated to receive much of that density. The redesignation of SLU to urban center meant the new neighborhood plan needed to provide for significant increases in density through land development (Department of Planning and Development 2005). Those increases were of course precisely the agenda of Vulcan and the City, who wanted to greatly increase biotech, condo, and commercial development. The increases were mostly threatening to low-income residents, who understood that the development agenda meant rising property values, and adequate affordable housing was not likely to be the central imperative of that agenda. Moreover, the old neighborhood plan was the work of the SLU community as it existed before the recent gentrification. That version of community included residents, non-profits, and businesses, with a much greater role for CNC and low-income Cascade residents (SLUFAN did not yet exist in its current form). The new constitution of community in SLU is much more dominated by business interests and property owners committed to the Vulcan–City redevelopment agenda. The ability of CNC and social-service advocates to influence the agenda was minimized. So the neighborhood planning process, hailed by advocates as the diamond in the tiara of the democratic "Seattle Way," was one strategy mobilized to serve a distinctly neoliberal agenda in South Lake Union.

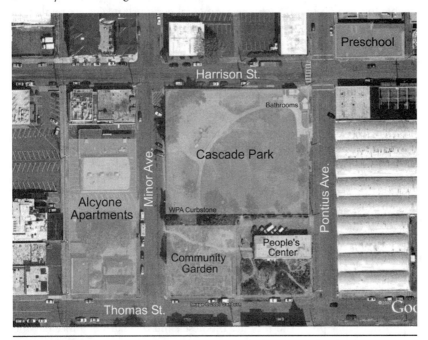

Figure 4.6 Cascade Park and the Alcyone Apartments (Source for satellite image: Google Earth)

In many ways, then, SLU is a stark case of the hegemony of neoliberalization. But it is worth remembering that hegemony is never total, that neoliberalization cannot completely erase what has come before. Rather it is layered on top of existing habits and structures, and the result is a complex mixture of political-economic norms and spatial outcomes. There is a park in Cascade that illustrates this complexity well (see Figures 4.5 and 4.6). Before the recent gentrification began, the park had been neglected by the city. Its facilities were in disrepair. It was mostly used by homeless people and people who used the social services in the neighborhood. The bathrooms in the park were commonly used for drug injection. One of Vulcan's first developments in the neighborhood was the Alcyone apartments across the street from the park. The apartments are geared toward young professionals who want an exciting, urban neighborhood. The prices mean that only fairly affluent people will live there: about $1,200 a month for a studio and $2,000 a month for a two-bedroom. Not wanting a run-down park across the street from their new development, Vulcan joined with other firms with interests in the neighborhood to fund a comprehensive redevelopment of the park. Their contribution of $600,000 was supplemented by $515,000 from the City. The result was a clean, well-ordered new space, complete with high-end play equipment for kids, a baseball field, a basketball hoop, and rejuvenated bathrooms. What Mike Davis (1990) has memorably termed "bum-proof" benches were installed so homeless people would be unable to sleep on them. Such environmental design strategies are reinforced by municipal ordinances prohibiting "camping" in public parks (i.e. sleeping in the park when it is closed, between 11:30pm and 4am). The idea of redeveloping the park, of course, was for Alcyone and other new developments to be much more attractive to people with kids. The funding was semi-privatized and the park is clearly designed to be an amenity that will attract tenants to the Alcyone. The building's promotional material features the park and its rejuvenation prominently.

However, the park was also funded in part by the City through a parks levy. It is an amenity for Vulcan, but it is also still a public park. In the corner nearest Alcyone, there remains an old concrete curb built during the depression by the most Keynesian of all entities, Roosevelt's Works Progress Administration. The WPA stamp and the year remain clearly visible, if one looks for them. Just south of that curb in the property adjacent to the park is a small but vibrant community garden, a shared resource among neighborhood residents that develops innovative sustainable gardening practices. Just to the east of the garden is the Cascade People's Center, a mostly volunteer, community-based non-profit organization that offers neighborhood residents a range of programs, including youth tutoring, early childhood education, a single-mom support group, a project to produce biodiesel, and the like. On a nice day the park is swarming with kids from the nearby pre-school, but homeless men and clients of neighborhood social services also still use the park to rest and socialize. While it is no longer possible to sleep on the benches, it is still possible to sit in the warm sun along the east wall of the park, or talk around the several picnic tables on the north side. Neoliberalization has greatly transformed the park—Vulcan and the City have leveled the field, installed new equipment, and renewed the entryways. But it has not entirely erased other uses, other people, and other values from the landscape. There remain, at least for the moment, alternatives to neoliberalization—still physically present in the space of the neighborhood—that can serve as the basis for resistance.

That resistance, in the case of SLU, has so far been latent rather than overt. But it is still worth thinking through what kind of resistance democratic movements for the right to the city might pursue, as a way to imagine responses to this worst-case scenario in which neoliberalization is largely dominant. In terms of discourse, the right to the city strongly resists the neoliberal narratives that construct SLU as empty. Of course SLU is *not* empty of inhabitants, of people who use the neighborhood space every day. The right to the city stresses that residents and the users of social services in SLU actively *inhabit* the area. They both depend on and intimately understand SLU's space in a way Vulcan, the City, and the biotechnology firms never could. As the *users* of South Lake Union, Cascade residents can claim a right to full and complete usage of its space. Their right to inhabit stands starkly against gentrification. Moreover, they can claim a right to full participation in the decisions that shape SLU. SLUFAN, the business-led community group, cannot claim the same right to SLU, as most of its members are property owners rather than inhabitants. CNC, on the other hand, is made up primarily of inhabitants, few of whom own property in SLU. The right to the city would claim the need to entirely *reverse* the current prominence of SLUFAN and marginalization of CNC in decisions concerning the redevelopment.

In terms of the democratic attitudes I advocate, claiming the right to the city would involve a range of groups mobilizing in coalition to make claims about the redevelopment of SLU. Those groups would share an equivalent, not identical, concern with inhabitance. SLU residents might be directly concerned with continuing their routines of inhabitance and protecting the particular spaces they most depend on every day. Social service agencies might be similarly concerned about being priced out of their offices, but they might also claim their clients' right to use the space near their office. The users of the community garden might also claim the right to continue their use and cultivation of that space. Together these groups would mobilize to claim a right to participate centrally in the decisions that shape redevelopment. They might not oppose biotech redevelopment entirely, but they would demand that any redevelopment meet the needs of existing inhabitants. That demand would necessitate that sufficient affordable housing was at least preserved if not built new, that redevelopment produced neighborhood spaces that met the needs of social-service clients and homeless people in a meaningful way, that existing collective relationships of care, centered on particular institutions like the community garden or the People's Center, were not disrupted, and that the redevelopment produced good jobs that offered opportunities to SLU inhabitants.

Now of course in this case, as in any case of gentrification, the makeup of the population of inhabitants is changing. New, wealthier residents are moving into the area and inhabiting it alongside the less-wealthy original inhabitants. The Alcyone is currently leasing, two other large (150–250 unit) buildings are just opening, and three others are designed and scheduled to be built in the next few years. New hip bars and gourmet delis are popping up every day. If a democratic movement were to construct a concept of equivalence around the right to the city in a broad, generic way, as an equivalent interest in the right of any inhabitant to inhabit, then the network would include the residents of the new developments, who also very much inhabit SLU. That broader construction of equivalence, of course, would make the movement more inclusive. But it would also introduce more instability, as those threatened with displacement would have a very different view of redevelopment than those whose housing tenure was secure. However, equivalence could also be constructed more narrowly, as an equivalent concern to claim the right to inhabit for those whose inhabitance is *threatened* by the gentrification of SLU. In that case, the network of equivalence would be defined more narrowly; it would be less inclusive but more stable. In the best case, it might also be possible to construct a network that included the new residents but defined inhabitance in the more narrow sense. How precisely to construct equivalence is a strategic political question. That is the flexibility of the right to the city and the radical democratic approach. Inhabitance as equivalence cannot be worked out before the fact. Democratic attitudes in concert with the right to the city offer some organizing rubrics, but they leave much room for political context. The question of how to define inhabitance and how broad to make the network must be worked out politically.

Another question that would confront any such movement is that of scale. While people could mobilize at the scale of the neighborhood as I have sketched above, they could also mobilize at larger scales as well. Certainly we might imagine a global democratic movement for the right to the city, but in terms of the SLU redevelopment, a citywide or even region-wide movement is a more immediate possibility. Groups from across the city who are concerned with the right to inhabit space could join the groups I have already mentioned to advocate for more affordable housing, for spaces that meet the needs of vulnerable inhabitants, for living-wage jobs for neighborhood residents. Moreover, such a citywide movement could call into question the public investment for infrastructure the City is currently planning. The claim could be that large amounts of public money are flowing into one neighborhood rather than being distributed more evenly throughout Seattle. Public investment is being used to promote economic growth that raises exchange values but does not secure use value for Seattle inhabitants. A claim to the right to the city could argue that public investment should flow to areas where inhabitants' needs are greatest, not where economic growth potential is highest. People in neighborhoods like Columbia City, Georgetown, Delridge, South Park, and Rainier Beach have argued that they have long-standing infrastructure needs (streets, sidewalks, parks) that the City has ignored, and yet it is committing huge amounts to redevelop South Lake Union. Moreover, that public investment is not likely to meet the use-value needs of SLU inhabitants (and indeed will threaten them), but benefit the exchange value interests of private firms. The right to the city would claim that such public investment should benefit those who inhabit the city, not those who own its property. The equivalent agenda here could be constructed as something like "public spending for use-value first." Under that banner, there is the potential that citywide movements for affordable housing, living-wage jobs, environmental justice, and community development could ally with more local neighborhood groups to make claims on redevelopments like SLU. Those claims could be to both particular, inhabitant-friendly development outcomes and a central role for the movement in decision-making. In the SLU case, a viable movement has not yet developed along these lines. However, the potential is there. There already exists significant citywide organization on issues like jobs, housing, and displacement, and there is a long tradition of neighborhood organizations advocating for their residents. While constructing such a network of equivalence around the right to the city would require significant effort, it is a perfectly realistic agenda, and it is worth trying.

The Seattle Waterfront

Another case from Seattle is less strongly marked by neoliberal hegemony. In fact, the waterfront process is much more typical of the Seattle Way and its deliberative commitment to public participation. However, the case raises some important questions about that deliberative-democratic approach to decision-making innovation in cities. Seattle's downtown has a large waterfront on its west side. Currently a large, elevated state highway (commonly called the "viaduct") runs along that waterfront between the skyscrapers and the water (see Figures 4.2 and 4.7). The 2001 earthquake in Seattle damaged the structure, and even though it was retrofitted there is consensus among civil engineers that the structure is unsafe and must be replaced, either by another elevated highway or some alternative (Post-Intelligencer Staff 2007). Because the viaduct is such a dominant feature of the downtown waterfront landscape, its impending demise has prompted an initiative to comprehensively reimagine and redevelop the area.

Figure 4.7 Seattle waterfront showing viaduct and downtown (Source: Washington State Department of Transportation)

The City of Seattle sees the current moment, according to their Central Waterfront Plan, as "an opportunity to develop a new 'Front Porch' that welcomes all ... and that represents an opportunity for economic growth for the city and the region" (Department of Planning and Development 2006a, p. 1).

To begin to imagine what a redeveloped waterfront might be like, in early 2004 the City convened a series of fairly innovative public events. They were labeled "visioning charrettes," and they brought together over 300 people in an intensive two-day workshop to produce design visions for the waterfront. A variety of people participated in the event, but as one might expect, professionals from the fields of architecture, planning, urban design, landscape architecture, engineering, and similar disciplines were a dominant presence both in terms of numbers and influence. Those professionals were joined by some community advocates and interested members of the public. The process was heavily influenced by deliberative-democratic ideas and, more specifically, communicative planning practice that draws inspiration directly from Habermas' theory of communicative action. Participants were divided into 22 teams, and each team developed their own vision for the waterfront. The teams sat together at round tables to deliberate toward a shared vision of what they thought the waterfront should be like. The idea in such charrettes is for multiple visions to converge into a new vision greater than the sum of its parts. It is inspired by the deliberative-democratic assertion that through cooperative communication, win–win solutions can be forged that could never have been imagined in a more conflictual political culture (Department of Planning and Development 2006b). The difference of the charrette model is that visual communication (of landscapes, land-use plans, etc.) is equally as important as verbal and written communication in the process of deliberation.

About two months after the visioning charrettes, the City held a Charrette Presentation and Exhibit at which the teams presented their visions to the general public for comments. Then, the City's planning department was charged with forging a Central Waterfront Concept Plan that would meld the various visions together into a unified document. The goal was for planners to use the charrettes as substantive content that they would translate into an official planning document that met legal requirements and professional standards. In order to ensure that the planners remained true to the charrettes, the City appointed a "Waterfront Advisory Team" (WAT), made up of various stakeholders (Washington State Ferries, property developers, the downtown art museum, labor, environmentalists, etc.) and design/planning professionals that supervised the planners' work over the next two or so years. The WAT meetings with the planners were very cordial and cooperative. The team members generally knew the planners and were keen to maintain a friendly relationship. It was also clear in those sessions that the planners were sincere in their desire to remain faithful to the charrettes. Often planning professionals exude a sense that their professional expertise should trump the uninformed, lay visions of the public. The planners in this case, on the contrary, saw their job as faithfully translating the charrettes into a workable plan. So the WAT's "watchdog" role was really more that of a friendly advisor offering helpful suggestions. In addition to the WAT playing an oversight role, in early-to-mid-2005 the City held several "open house" events across the city to report progress to the public and to take public feedback. Those who attended these sessions were mostly planning and design professionals, but there were some interested citizens as well. The designs presented were impressive graphically but were quite vague. It was difficult to offer substantive feedback. In many respects, the open houses were more pageants to publicize and celebrate the public process, rather than a chance to get feedback that would substantially alter the direction of the plan. As with the WAT meetings, the mood at these events was distinctly festive and collaborative, rather than tense and confrontational. Once those public events were completed, the planners, working with the oversight group, finalized their plan. It was unveiled at yet another public event in mid-2006, and the mayor approved the final Waterfront Concept Plan and forwarded it to the City Council.

However, in the summer of that year, the deeply connected issue of the viaduct began to get quite muddled. While there was consensus that the existing structure was unsafe and needed to be removed, there was much debate about what would replace it. One option was to rebuild a similar, but larger, elevated highway. Another option was to replace the viaduct with a tunnel underneath the waterfront area. The mayor and most of the council agreed that the tunnel was the best way forward. One reason for that position was that opposition to a rebuilt viaduct is widespread and visceral: most people see the viaduct as an ugly, noisy blight on the waterfront and downtown. It inspires the kind of hatred that could turn into electoral backlash should decision-makers decide to rebuild it. Moreover, the viaduct renders the wide swath of waterfront land underneath it virtually unusable. A tunnel, by contrast, would create large open spaces on the surface where the viaduct once stood, a more-or-less blank slate that could be used in a variety of ways. Most of the visions for the new waterfront developed at the charrettes and in the plan imagine some variation of sun-swept promenades (in Seattle!) and open space along an azure bay. Those visions depend on a tunnel, not a new viaduct.

The main problem with the tunnel was that it was the more expensive option (cost estimates were roughly $2.5 billion for the viaduct and $3.5 billion for the tunnel (Lange 2006)). That expense is a problem because the viaduct is part of a Washington State highway. Therefore, the cost of any project would be borne primarily by taxpayers across the state (in addition to Federal taxpayers, since Federal grants would help fund the project). But taxpayers in Eastern Washington are strongly opposed to paying for the needs of Seattle, especially when Seattle is proposing the most expensive option. Perhaps more so than in any other state in the country, Washington is split between the urban, educated, industrial, wealthier, progressive metropolis and its rural, less-educated, farming, poorer, and conservative hinterland. Representatives from Eastern Washington in the state congress have made funding for a tunnel a very difficult proposition. Governor Christine Gregoire, who won her office in an election that was so close it dragged on through several recounts and legal challenges, cannot afford to alienate rural voters, and she has opposed the tunnel. Stoking that opposition is the infamy of Boston's "big dig," a tunnel that has experienced all sorts of funding overruns, deadly engineering flaws, and shoddy construction. Therefore, despite consensus on the tunnel among local decision-makers, its future remains very much in doubt. Without a clear decision on the transportation plan, it is difficult to construct a serious land-use plan for the waterfront, since the two highway options offer very different environments. So, the Waterfront Concept Plan is currently in limbo, awaiting the outcome of the struggle over transportation options. Recently, Seattle city voters voiced their opinion in a special advisory vote; they rejected both the tunnel and the viaduct. That vote further muddied the waters, and it gave momentum to a third option, that of removing the viaduct in favor of improved surface streets and augmented public transit. I elaborate that third option in more detail, below.

Whatever the transportation outcome, it is quite clear that the waterfront plan as it stands was produced by a far more democratic process than the City pursued in South Lake Union. The charrettes and the clear commitment to public input on the part of planners are quite different from the active suppression of meaningful public debate in SLU. Nevertheless, there were important democratic problems with the waterfront process. First, the actual participants were far from representative of the city's public. A significant proportion were professionals with potential material interest in the outcome of the process. One charrette team, for example, was almost entirely made up of employees of the multinational design firm EDAW. EDAW has been contracted to do similar projects in cities like Boston, Los Angeles, and San Francisco, and it is no stretch to say that it participated in the charrettes to get its foot in the door for any future redevelopment design contracts. Most other participants were in some way stakeholders in the specific area of the waterfront, rather than citizens concerned in general about their city's "front porch." Such direct stakeholders included, for example, the Downtown Seattle Association, the chamber of commerce for the downtown area. So the process had distinct problems with inclusiveness. Habermasian ideals demand that all affected parties are at the table for deliberation. The charrettes clearly failed to achieve that inclusiveness, and they therefore failed to achieve the deliberative-democratic model on its own terms. Critics of the deliberative approach make the case that truly inclusive processes are a conceptual impossibility, that the appearance of inclusiveness will always mask the exclusion that is impossible to avoid, and that decisions are therefore likely to favor the included over the excluded. However, in this case there was quite a lot of room for improvement in terms of inclusiveness before that more subtle critique would need to be mobilized. Even the organizers lamented the predominance of professionals and the absence of average citizens.

However, in another sense the charrettes did meet deliberative standards. Most deliberative processes, as actually realized in practice, follow a stakeholder model. While Habermas would demand that *anyone* affected by the decision be defined as a stakeholder, in common practice stakeholders are defined as those who have a particularly *significant* stake in a decision. In that way, the stakeholder-heavy profile of the participants was quite in line with typical deliberative democratic practice, even if it did not match the ideal. Stakeholders had a prominent say in the process, both because they participated and voiced their opinion, and because planners assumed explicitly that they needed to please all significant stakeholders for their plan to be successful. The problem with that slippage between ideal and practice is that it diminishes the influence of liberal-democratic citizens (the "public" concerned about their front porch) and increases the influence of other interests. Those other interests are diverse, but one group is nearly always defined as a stakeholder: large landowners. Almost without fail, any large landowner affected by a particular decision will be represented as a stakeholder in a deliberative process. But very often, as in the waterfront case, the civically-minded citizen, interested only in that she is concerned about the fate of the front porch, is not well-represented in such processes. In the waterfront case, the stakeholder profile did have some breadth. However, as usual, it was careful to include significant landowners and large businesses that operate in the area, such as, for example, Harbor Properties, a developer of commercial and residential property, which has significant holdings in the waterfront area.

To reiterate, the approach of formally including property owners among many other stakeholders is a far cry from the closed-off model of SLU, where the principal owner and the City make decisions virtually alone. Nevertheless, such deliberative practices also favor a neoliberal agenda, in more subtle but perhaps more dangerous ways. The waterfront process explicitly included property owners *just because they are property owners*. For owners, such formal inclusion is an improvement over a traditional liberal-democratic model that excludes property owners because they are not citizens. Of course, property owners have always found ways around or through liberal-democratic structures. Nevertheless, the deliberative alternative offers them a chance to operate out in the open, as a legitimate democratic participant. In the deliberative model, the formal inclusion of stakeholders is usually paired with the principle that all stakeholders must be satisfied with a decision (Innes 2004). Stakeholders can't expect to get everything they want, but they can expect to get their most essential needs met, to be able to "live with" the decision. Planners in this case were clearly concerned about stakeholder satisfaction, as we saw above. They continually worried about how key landowners (whether private firms or public agencies) would react to a given idea being considered. So, if property owners are formally included as stakeholders, and if all stakeholders must be satisfied with the decision, then the process ensures that property owners will get what they need. Therefore, deliberative processes like the waterfront formally protect the interests of property owners, and the open, public process means the eventual decision is seen as highly democratic. Such processes allow property owners to get what they need while also getting the very important added benefit of political legitimacy. For a given project, such legitimacy minimizes uncertainty (legal challenges, organized resistance, etc.), which is an extremely valuable commodity for large-project development. More generally, such legitimacy helps manage the democratic deficits that neoliberalism tends to produce. If managed skillfully, deliberative processes offer the neoliberal agenda a very attractive political fix for their ongoing crises of democratic legitimacy. In other words, to put the point more provocatively, over the long term neoliberalization is almost surely better served by a waterfront model than by an SLU model.

On top of those factors, deliberative processes like the waterfront offer aid to neoliberalism because they do not dare to rethink some basic assumptions of the neoliberal world view. In the waterfront case, much of the area is owned by public entities (e.g. the City, the County, the Port of Seattle). The rest is privately held. The planners felt they had wide latitude to plan the public land, but little influence over the private holdings, other than typical tools like zoning and incentives. Existing regimes of property ownership and property rights, in other words, greatly constrained any vision the charrettes produced. While most of the land adjacent to the waterfront is owned by public entities, most of the rest is owned privately (see Figure 4.8). That makes it difficult to carve out swaths of land for large-scale parks, promenades, and gathering places, all common features of the charrette visions. Moreover, the public entities all operate independently, and, like the private owners, they are keenly concerned with the exchange value of their holdings, since many of them are portfolio investments for the public fisc. Turning their investments into parks (rather than, say, condominiums) would devalue their portfolio. The planners could see those constraints, and they spent considerable time thinking about the tools they needed to overcome it. The best possibility, one they very much hoped for, was a redevelopment authority with the power of eminent domain. In Washington State, however, such authorities are heavily circumscribed by law. So the planners were really quite limited in their ability to develop a truly comprehensive rethinking of the area. The charrette process, no matter how democratic it was, eventually ran up against the age-old barriers of property rights and the supremacy of exchange value.

Moreover, in this case there was minimal public funding committed to make the plan a reality. The planners were developing a plan to make the public's vision real, but there was no money to actually *do* what the plan called for. That lack of public funding is in part due to the more general problem of the fiscal squeeze that all cities face. However, as we saw in SLU, when a city sees an economic growth opportunity it is more than willing to commit significant resources to make it happen. The waterfront redevelopment as it emerged from the charrettes was much more a vision for reclaiming the waterfront for users, to create a civic meeting place for mixing and exchange. Such civic revitalization does offer some economic development potential, in the sense that it helps produce a general image of a lively and engaged city with an amenity-strewn "front porch." Such signature spaces are of course an important element of the package that cities present to potential investors. However, in this case the City clearly thinks SLU is the big fish here in terms of economic development. They are not willing to commit the same kinds of resources to the waterfront. Therefore, planners were forced to imagine creative ideas to fund redevelopment. They worried

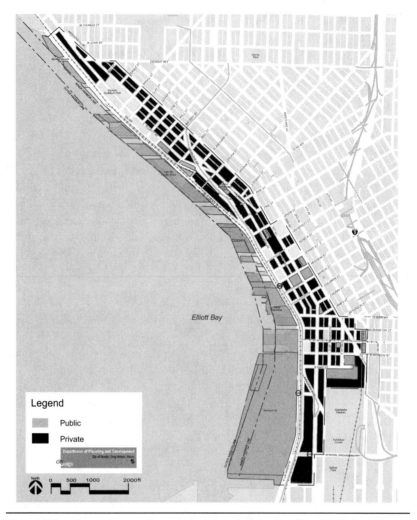

Figure 4.8 Map of property ownership on Seattle's waterfront (Source: City of Seattle, Department of Planning and Development)

that the most likely scenario would be to create public–private partnerships whereby development firms would be enticed to invest capital to redevelop the area. Of course, such a structure, typical of neoliberal governance, would require a return on capital investment. It would require that exchange value play a prominent role in shaping the redevelopment. Planners saw clearly that such a structure would significantly threaten, if not completely alter, the visions that came out of the charrettes. So again, however democratic the charrette process was, it existed inside a wider neoliberal political economy that severely constrained what it was able to achieve.

People's Waterfront Coalition

There is one other part of the story that is important to explore. Mainstream decision-makers have assumed that there are only two transportation options for the waterfront: rebuild the viaduct or construct a tunnel. Those choices are supported by an orthodox assumption in transportation politics and engineering: any lost capacity in the transportation infrastructure must be replaced. Traffic congestion is best addressed by maintaining or increasing capacity. The question of traffic is an especially sensitive one in Seattle, where congestion is among the worst in the country. Traffic is a significant competitiveness issue. Boeing has repeatedly complained that they cannot move goods efficiently through the region, and they have threatened to move production somewhere with better flow. The Washington Competitiveness Council found improving transportation infrastructure to be the most important competitiveness issue. Moreover, the worst bottleneck in the region is the I-5 corridor downtown. For vehicles going through downtown, the only alternative to I-5 is the viaduct (see Figure 4.2). So any decrease in the capacity of the viaduct artery, the orthodox view goes, would further clog I-5. Given those assumptions, the overwhelming conclusion is that when the old viaduct is torn down, it must be replaced with at least the same if not more capacity. Both the tunnel and a new viaduct would at least preserve the existing capacity.

However, an alternative assumption, one favored by a fairly small minority of planners and urban designers, is that reducing capacity can, paradoxically, help reduce congestion. The long-term solution, the argument goes, is not to remain on the capacity treadmill whereby more infrastructure encourages more drivers which requires more infrastructure, etc. Such an approach only preserves the problem, and worse, it exacerbates the growing problem of carbon emissions and their impact on climate change. The solution, instead, is to reduce capacity, to squeeze drivers out of their cars and push them into mass transit. It is more mass transit and fewer vehicles, not increased capacity, that will solve the congestion and emissions problem. It will give travelers real transportation options, and improve the overall efficiency of the system. Following that alternative argument, one group in the charrette advocated not a tunnel or a rebuilt viaduct, but no highway at all. They favored instead a surface-grade boulevard whose goal was not primarily to move cars but to be a cohesive part of a reimagined waterfront area. The plan included all the civic spaces and pedestrian corridors of the other plans, but it earned that space by eliminating the highway, not putting it underground. When City decision-makers made it clear that they favored a tunnel, members of the charrette group formed what they called the People's Waterfront Coalition (PWC) to build a groundswell of opposition to building any highway along the shore. Initially the coalition had limited success, since the idea of not replacing capacity was so distinctly outside the pale. Recently, however, as I mention above, the tunnel plan has encountered significant difficulty. That difficulty is mostly borne of larger-scale political factors unrelated to the PWC's efforts. If the tunnel fails to go forward, few are excited about a rebuilt viaduct as an alternative. There is therefore a potential gap that the PWC's no-highway plan could fill. Proponents of both the tunnel and the viaduct are beginning to favor the no-highway option as their second choice (Garber 2007). And so an opportunity has opened for the coalition to press its agenda with those inside government who are becoming increasingly sympathetic to the no-highway plan. Of course, the voters' recent rejection of both the tunnel and viaduct only widened that opportunity.

The leaders of the PWC are more planners and designers than they are organizers and activists. And they are well-connected to decision-makers. So their actual strategy has been less a popular groundswell built through grassroots organizing than an effort at networking and politicking among decision-makers. Their agenda is not so much to democratize decision-making as to undermine an orthodoxy and advocate for a marginalized idea in land-use and transportation planning. Nevertheless, PWC has opted out of the deliberative process in place and pursued instead a quasi-social-movement approach, one that sees struggle and argument as the stuff of politics. PWC takes an agonistic approach to their politics. Their goal is to resist the current hegemony in planning thought and work toward a new hegemony, and to do so through mobilization, not deliberative consensus. In that respect, then, they exhibit some of the attitudes and methods associated with radical democracy (as opposed to liberal or deliberative democracy). However, it is fair to say that to date the movement is one of planning elites engaging agonistically with other planning elites. They are more concerned to change the way planners think about planning than they are to democratize planning decisions. But even if the PWC is not properly a movement for radical democracy, they do teach us important lessons by proxy. They illustrate how deliberative processes like the charrettes are not forums in which orthodox assumptions are likely to be effectively challenged (in this case, assumptions about traffic capacity). They suggest that movement politics are a better way to identify and challenge the constraints of orthodoxy. In the case of the PWC, that strategy may or may not bear fruit. If it does, it will not be the PWC alone that has forced decision-makers to reexamine their assumptions, but a confluence of many political factors, including the PWC. However, what is instructive is that the PWC felt a movement strategy was their best option. Inside the confines of the deliberative process, they found almost no headway was possible. Their experience suggests that the best option for counter-hegemonic projects is agonistic struggle. If sufficient opportunity presents itself in the political landscape, or, more accurately, if the movement can creatively discover and exploit that opportunity, then there are real possibilities for challenging the status quo and making progress toward counter-hegemonies. The case of the Duwamish River

Cleanup Coalition, which I will explore next, melds this counter-hegemonic politics of opportunity with a more explicitly democratic agenda.

The Duwamish River Cleanup

The Duwamish River is Seattle's main river. It is formed by the confluence of the Green and Black Rivers, and together the three rivers drain a large watershed south and east of downtown Seattle that begins in the Cascade Mountains and ends up in Elliott Bay, the body of water just west of downtown Seattle (see Figure 4.9). The Duwamish runs entirely through urbanized areas, although the Green begins in the wilderness of the Cascades. The last several miles of the river are channelized, and the most downriver stretch of the Duwamish is the site of the largest concentration of industrial activity in the Puget Sound region. The bulk of that industrial activity, including such uses as cement production, food processing, aerospace manufacturing, paper and metals fabrication, and boat building and repair, is inside the Seattle city limits. That stretch of the river is also the site of the Port of Seattle, one of the largest ports on the west coast. As one might expect with a heavily industrialized river, the lower Duwamish is extremely polluted. Chemical pollutants from industry include high levels of hazardous materials in water, sediment, and soil near the river, including polychlorinated biphenyls (PCBs), polyaromatic hydrocarbons (PAHs), mercury and other metals, and phthalates. In addition, over 100 storm drains carry a variety of pollutants from pavement into the river. More periodic but more destructive are the combined sewer overflows (CSOs) that dump untreated sewage into the river when rains have been heavy and sustained enough to overload the city's sewage system. As a result of that pollution, the United States Environmental Protection Agency listed the final five-mile stretch of the Duwamish as a federal Superfund site on September 13, 2001.

The Superfund designation sets in motion a complex mixture of federal, state, and local procedures that are shaped to a significant degree by the political, economic, and cultural specifics of the Seattle context. The national-scale Superfund regulation is based largely in a law called CERCLA, or the Comprehensive Environmental Response, Compensation, and Liability Act. The original idea of Superfund was that the federal government would tax chemical and petroleum companies to build up a "Superfund" that would pay for environmental cleanup in places such companies had been primarily responsible for polluting. In addition, the law put polluters on the hook: where parties responsible for the pollution can be identified, they must help fund the cleanup. That extremely assertive federal law was passed in December 1980, in the last weeks before Reagan became president. It was born, therefore, near the beginning of what is commonly considered the dawn of neoliberalism's ascendancy in the United States. However, the Nixon administration had

The Green/Duwamish River Watershed

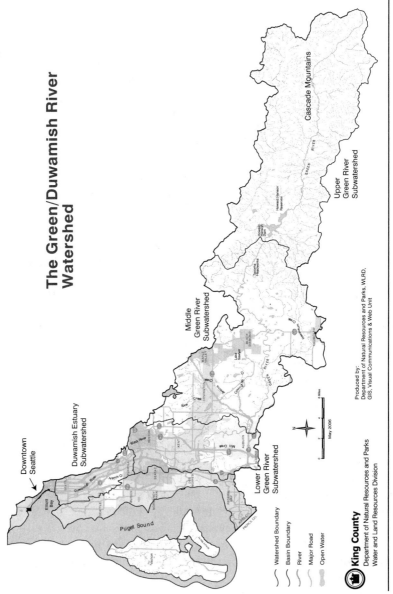

Figure 4.9 Green–Duwamish River watershed (Source: King County Department of Natural Resources and Parks)

already launched the beginnings of the neoliberal agenda earlier in the decade. Those initiatives actively undermined Keynesian federal programs, and, more specifically, programs that came out of Johnson's Great Society agenda. The governance of community development, for example, which in the 1960s had taken on all sorts of public participation provisions in the wake of the brutal urban renewal projects of the 1950s, was, under Nixon, increasingly devolved to local authorities, and the federal government withdrew significantly from both funding and oversight. Superfund, therefore, which created strong federal regulation and mitigation of the harm corporations do to the environment, can be seen on the one hand as a parting shot from a dying Keynesian tradition. On the other hand, however, the 1970s also saw a wave of laws that gave the federal government much greater authority to protect environmental well-being. The Environmental Protection Agency was created in 1970, and it was followed by the Clean Air Act (1970), the Safe Drinking Water Act (1974), the Toxic Substances Control Act (1976), the Resource Conservation and Recovery Act (1976), and the Clean Water Act (1977). All of these are managed and enforced by the EPA, and CERCLA joined that list in 1980. So in other respects, CERCLA resulted from a surge in the 1970s of what might be called "environmental Keynesianism" that was partly independent from the wider neoliberalization of Keynesian political-economic policy (as with community development).

As we might expect, however, such strong federal regulation of the environment has come under fire during the neoliberal era. The EPA in general, and CERCLA in particular, has been subject to neoliberal reimagining over the course of their history. One major shift has been greater devolution of authority over cleanups (Adler 1998; and see Schoenbrod 1996). Partly as a result of the neoliberal philosophy that big government should not dictate solutions to people and places, local agencies (both governmental and not) at the state, county, and city scale have taken on increasing responsibility for the everyday management of cleanups while the EPA increasingly plays an oversight role, ensuring that the provisions of CERCLA (and other laws) are met. So in that sense, CERCLA governance fits well the neoliberal pattern of devolving and outsourcing governance. But as with all such governance shifts, the federal–local relationship is highly malleable and context-specific. While the trend has been toward devolution, a whole constellation of different decision-making arrangements can be made. Within limits, each cleanup can be governed very differently from the next. They can have a different mix of agencies at different scales, and the authority each holds with respect to the others can vary widely as well. Recall that in general neoliberalization involves devolution and outsourcing such that governance is made increasingly flexible, usually in a way that benefits capital interests. That has been the case with Superfund as well: a proliferation of ad hoc and special purpose entities increasingly carries out the everyday decision-making in Superfund cleanups, while federal and state governments play mostly an oversight role. While we must remember that such "flexibilization" is designed to advance the neoliberal agenda, it can easily have a double edge. Increased openness and uncertainty in the governing regime tends to create political opportunities that social movements can exploit.

And in fact that exploitation has been a key part of the Duwamish case. For the Duwamish cleanup an extremely complex governance regime has emerged. The EPA and the Washington State Department of Ecology (usually called just "Ecology") have taken joint oversight for the cleanup. The agency actually carrying out the cleanup was created from scratch just for that purpose. It is a public–private entity called the Lower Duwamish Waterway Group (LDWG). LDWG is made up of four entities: the Boeing Company, the City of Seattle, King County, and the Port of Seattle (see Figure 4.10). Their selection was not random: under CERCLA, parties responsible for the pollution (called potentially responsible parties or PRPs and often pronounced "perps") are required to help pay for and sometimes conduct the cleanup. In this case, the four LDWG members were the four major PRPs who could be identified and were still in existence. PRPs can be designated in two main ways. First, there can be direct evidence they polluted the environment. Second, they can own the land and have assumed liability from a previous owner who polluted. Since the pollutants in the river and on its banks can last for years, very often the actual polluter went out of business long ago. In the Duwamish case, Boeing

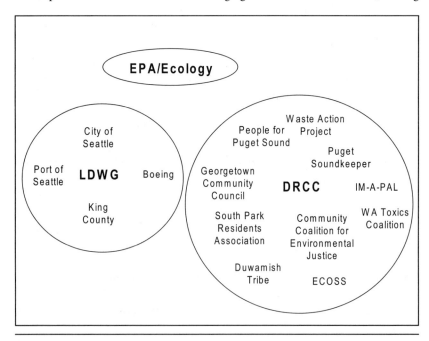

Figure 4.10 Governance structure of the Duwamish cleanup

is an identifiable polluter. King County and the City of Seattle are as well, because they are responsible for some of the combined sewer overflows on the river. The activities of the Port and their tenants also directly contributed to contamination. However, all four parties are also PRPs in the second sense— they own significant property in the Superfund site and so they have assumed liability through purchase.[1]

The story of LDWG's birth, like Superfund more generally, reveals the tension between the Keynesian and neoliberal tendencies that surrounds environmental regulation. Typically, when a Superfund site is declared, the EPA will identify PRPs and ask them to investigate pollution levels and carry out a cleanup. In this case, there was little doubt that the County, City, Port, and Boeing would be PRPs if a Superfund site was declared. Seeing the writing on the wall, the four PRPs attempted to create an alternative governance structure that would prevent Superfund regulation. They formed LDWG as a legal public–private partnership so that they could enter into a binding agreement with EPA to do the cleanup without a Superfund declaration. The agreement would thus preempt Superfund law and allow the PRPs to negotiate specialized terms that were more favorable to them. That "creative approach," they claimed, would be "more efficient and cost-effective than a Superfund listing" and would achieve cleanup "in a shorter amount of time with less process" (Port of Seattle 2000). They offered standard neoliberal logics about the inefficiency of state regulation and the superiority of "creative" new governance approaches. The agreement required the consent of each of the four members of LDWG, the EPA, the State Department of Ecology, and the trustees, which include the National Oceanic and Atmospheric Administration, the U.S. Fish and Wildlife Service, the Muckleshoot Tribe, and the Suquamish Tribe. The agreement fell apart, however, largely because Boeing was not willing to accept the trustees' demands to restore fish runs (Torvik 2000).

As a result of that failure, a Superfund site was declared in 2001. That declaration triggers all sorts of federal laws and regulations (e.g. it allows PRPs to be assigned formal legal liability). However, as we have seen, the Superfund designation does not mean a great leap backwards into full-blown Keynesianism. Superfund governance, like so much neoliberal governance, is best seen as a palimpsest. It is made up of neoliberal elements (e.g. devolution and public–private partnerships) that have been laid on top of old Keynesian and Great Society ones (e.g. federal imposition of legal liability on polluters). Sometimes the old Keynesian structures have been entirely erased, sometimes partly, sometimes not at all. As Superfund governance has been neoliberalized, it has become fairly typical EPA practice to turn over study and management of cleanup to PRPs. In urban areas, PRPs are often public agencies, because they tend to own lots of urban land. So it is not unusual to have both public and private PRPs. They often form some kind of public–private partnership to directly govern urban Superfund cleanups. That fact might be read as simply a manifestation of the wider neoliberalization of governance. Asking polluters to study the problem and carry out a cleanup seems a clear case of the fox guarding the henhouse. However, the kernel of this structure lies in the Keynesian roots of CERCLA: the federal government requires polluters to pay to clean up the mess they made. Any time the actual polluter cannot be identified, money from the "Superfund" must be used to pay for the cleanup, and the EPA wants to minimize such public funding as much as possible. So the structure intends to hold polluters financially accountable in a direct way. On the other hand, such accountability does not also mean they have to be given so much control over the cleanup process. While it is true that EPA–Ecology have final approval power, there is much room for LDWG to shape the extent and the cost of the cleanup. That shift in control, from the federal government to more local-scale and non-government entities, is of course characteristic of neoliberalization. Such neoliberalization raises the kind of red flags in terms liberal democracy that I examine in Chapter 1. In the Duwamish case, Boeing is formally unaccountable to the public, and the public agencies in LDWG are accountable only in a very remote way. That is, most decisions made in the name of the Port or the City or the County are made by their professional staff (planners, ecologists, engineers, etc.). The elected councils of those public agencies have ultimate veto power over staff decisions, but that power is rarely used (although it has been once in this case).

If that were the whole story, we might be tempted to join the chorus of the political-economy literature and file this case with South Lake Union as yet another example of the neoliberalization of urban politics. Even if we added the caveat that such neoliberalization is always incomplete, it seems the fox has taken control of the henhouse on the Duwamish. The federal government has turned over everyday control of the cleanup of a severely polluted river to an unelected, special-purpose, and quasi-public entity made up of the polluters themselves. However, there is another wrinkle to Superfund law that makes this case quite a lot more promising than that. In an echo of the Great Society's public participation initiatives of the 1960s, CERCLA includes explicit provisions for what the EPA calls "community involvement" in the Superfund process (Office of Emergency and Remedial Response 2005). The idea is to ensure that people who live and work near the cleanup site be involved in decisions about cleanup. Because it tends to define "community" in terms of geographical proximity, it constitutes a fairly narrow imagination of the idea. However, it is important to stress that there is much room for maneuver in Superfund's community involvement provisions. Once a site is listed for cleanup, members of the public can form what is called a Community Advisory Group (CAG). A CAG serves as the officially recognized liaison between the community, the EPA, and the entity conducting the cleanup. In other words, once a CAG is formed, it becomes the official "community" in the eyes of the EPA. In a sense, that structure can be seen as a way to tame public involvement, to make sure it conforms to predetermined parameters and does not spill over to become unmanageable resistance. My discussions with EPA officials support this notion in that while they themselves might personally support robust public involvement, as representatives of the EPA they are somewhat cautious about pushing any envelopes in terms of participatory practices. They are not overly optimistic about how receptive the agency more broadly would be to more vigorous public involvement. In general, it is largely true that Superfund law calls for public participation, but that participation is intended to be contained and manageable.

However, as I said, there is room to maneuver here. The participation provisions are not fine-grained enough to overly constrict CAGs if they are insistent, active, and imaginative. If CAGs claim access or rights or privileges, there is rarely an explicit prohibition in the existing law. So the EPA frequently has the option of granting or denying such participation claims. A sympathetic cadre of officials can allow robust participation if they choose to, or if they are convinced by the CAG to do so. Moreover, there are a range of very useful tools that the EPA makes available to CAGs. One resource available to CAGs is Technical Assistance Grants (TAGs), which allow the CAG to hire their own technical experts. Such expertise is critical because the first step in a Superfund cleanup is a series of scientific studies of the pollution problem. The entity carrying out the study must sample the site to determine pollution levels for a range of pollutants; they conduct ecological, hydrological, geological, atmospheric, chemical, and other analyses to determine what impact the pollution is having, how much should be removed, and how that removal should be done. Then they must plan and conduct highly technical engineering tasks to carry out the cleanup. Average citizens lack most of those specialized knowledges, and so they need to hire experts to help them evaluate and respond to the lead entity's proposals in a way that cannot be brushed aside as uninformed or unrealistic. Equally critical is the CAG's ability to maintain and manage itself. In the Duwamish case the CAG has funding from the Washington State Public Participation Grant Program, through the State Department of Ecology, to hire a full-time, experienced coordinator and pay for outreach activities. As a result, in this case the job of running the CAG is professional, not volunteer. Participating meaningfully in a cleanup as complex as the Duwamish requires more than just an activist's spare time. It requires a talented and full-time professional.

In addition to those tools, there are several other support structures in place for CAGs as well. Technical Outreach Services for Communities also provides technical assistance. The Superfund Job Training Initiative offers training programs that help community members acquire skills that allow them to participate in the cleanup and help them find employment once the cleanup is over. It is important to stress that each of those structures can be used or not, and used effectively or not. Much depends on the mobilization and commitment of the community, and how shrewdly they take advantage of (and expand) the structural opportunities the EPA and other entities offer. Where such community energy and expertise is missing, the structures, by themselves, rarely produce robust community involvement.

In the Duwamish case, "the community" has taken great advantage of what the EPA and Ecology offers. A group called the Duwamish River Cleanup Coalition (DRCC) formed, petitioned to be the CAG for the cleanup, and was accepted by the EPA. The DRCC is a broad coalition of environmental, resident, tribal, environmental justice, and small-business groups (see Figure 4.10). The environmental activist element is the largest cluster and includes People for Puget Sound and Puget Soundkeeper, which are made up of mostly middle- and upper-class Anglos. They are well-educated, have a long history of environmental advocacy, and are quite well connected with decision-makers at the local and state levels. Another environmental group, Washington Toxics Coalition, is a statewide organization working to reduce human exposure to toxic pollution. Waste Action Project is a public interest group that uses primarily lawsuits to ensure enforcement of environmental laws. The IM-A-PAL Foundation is the creation of John Beal, a well-known activist who restores creeks and habitats in the Duwamish basin to greater ecological health.

The environmental groups are joined in DRCC by two neighborhood-resident groups, the South Park Residents Association and the Georgetown Community Council. South Park and Georgetown are the two residential neighborhoods most affected by the sites in the cleanup (see Figure 4.11). Their residents are relatively low-income and non-white compared with the rest of the city. According to the 2000 U.S. Census, residents in Seattle are 74 percent white; residents in South Park are only 34 percent white and in Georgetown 56 percent. South Park has a notable concentration of Hispanic residents: 37 percent (compared with 5.3 percent citywide). While the median household income for the city is $49,469, for South Park it is $30,917 and for Georgetown it is $33,654. Not surprisingly, the two neighborhoods—in close proximity to a range of industrial land uses and relatively poor and non-white—are disproportionately burdened with environmental hazards like air pollution (e.g. cement and paint industries in South Park), water pollution (e.g. the river, as well as plumes of toxins in the groundwater under Georgetown), and soil contamination (e.g. PCBs in road soil, public rights of way, and residential yards in South Park). Partly because of the environmental justice issues in South Park and Georgetown, another member of the DRCC is the Community Coalition for Environmental Justice, itself a coalition of various activists across the city who campaign against environmental injustice, displacement, and racism.

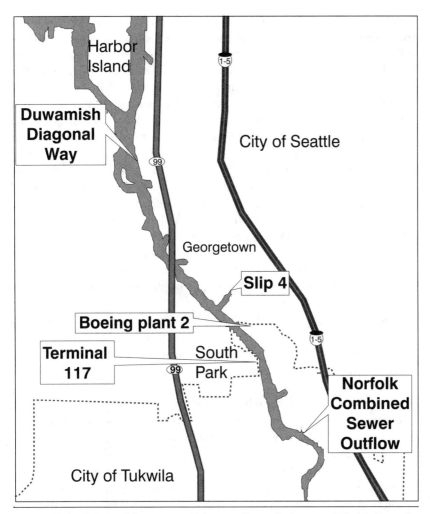

Figure 4.11 South Park and Georgetown, with hotspots (Source for base map: EPA, http://yosemite.epa.gov/R10/CLEANUP.NSF/ldw/site+map)

Another member of the DRCC is the Duwamish Tribe. The Duwamish are the people of Chief Si'ahl (Seattle) who inhabited the lower Duwamish watershed (as well as parts of the Cedar and Black watersheds) at the time of the first white expeditions to the area in 1851. They are not a federally recognized tribe, but rather a 501(c)3 non-profit corporation. That distinction is important because Superfund law makes significant provisions for federally recognized Native American tribes, but those do not apply for the Duwamish. Nevertheless, the Duwamish have extensive interests in the lower Duwamish and in the cleanup. The river and its wider system are part of the traditional fishing grounds of the Duwamish, as well as the Suquamish and Muckleshoot tribes. The Suquamish and Muckleshoot, because they have federal treaty rights, still engage in commercial fishing. And subsistence fishing continues on the river, both among native populations and non-native ones (Washington State Department of Health 2003). Another interest for the Duwamish is cultural and historical. Many sites along the river are of great archeological importance to the tribe. Such cultural memory is critical for a group whose continued existence as a people is at stake. Moreover, that cultural memory is bound up tightly with the ecological health of the river. Before white conquest, the Duwamish lived in and relied on the Duwamish Valley. The industrialization of the Valley and the progressive destruction of the river's ecology is intimately connected to the assault on the Duwamish tribe's way of life. In many ways, a claim to restore the river's ecological health is also a claim to redress the injustice of white conquest. For the Duwamish, it is difficult to separate out their cultural, historical, economic, and environmental claims. They want, in some sense, a reparation: a renewed commitment to the idea that the Duwamish is not just an industrial waterway but a complex riverine ecosystem that should be preserved, restored, and valued for its own sake.

One last member of the DRCC is the Environmental Coalition of South Seattle (ECOSS), an association of small businesses in the Duwamish area that advocates for their interests, with a particular focus on helping them develop sustainable environmental practices. The DRCC, then, is a coalition of coalitions. They do not define "the community" to be only those in proximity to the cleanup. Rather they bring together many different cross-cutting interests—environmental, cultural, resident, small-business, health, justice— into what we might call a network of equivalence. The DRCC is not a network of only inhabitants. That is, many in the network do not themselves inhabit the watershed. However, the equivalence that the network shares is an agenda to protect and restore inhabitance in the watershed. They share a commitment to the river and its watershed as habitat, as a complex place that is inhabited by humans and other species. Given the nature of their coalition, the DRCC is careful to very clearly include both humans and non-human species. While human health and well-being is central, it does not necessarily trump, nor is it necessarily separable from, the well-being of the river's ecological system. That kind of shared commitment can be difficult to sustain. However, despite the coalition's distinct diversity of interests, so far it has been quite cohesive. It makes decisions by the agreement of all members, so that dissent by any one member can potentially prevent them going forward. While they do have provisions for making decisions by majority vote if necessary, they have not yet failed to reach agreement on a decision.

The DRCC's agenda of inhabitance stands in contrast, broadly speaking, to the agenda of LDWG, the public–private partnership. In thinking about the contrast, it is instructive to begin by examining the different names of the groups. DRCC refers to the Duwamish as a river; LDWG calls it a waterway. The lower Duwamish was straightened and deepened years ago to accommodate the Port and the industrial uses, and the straightened stretch has long been referred to as a "waterway." A waterway is a transportation corridor that is functional for an industrial economy. A river, of course, is something else, something more explicitly ecological. A waterway is infrastructural; a river is inhabited. The members of LDWG are diverse and have many different interests, but an overarching commonality is that (1) they are funding the cleanup and so have an interest in minimizing cost, and (2) they are major landowners in the area and so have an interest in the property values along the river. Those two goals sometimes conflict, of course, since more comprehensive cleanup is more costly but results in more saleable land. But the predominant concern from a property owner's perspective is the line between saleable and non-saleable. For a polluted property that cannot be legally sold on the market, one must reduce pollutants to a level that is below a legal maximum, and then the property can return to the market. Of course, in a cleanup of this scope there is much more one can do than merely reduce pollution to below legal maximums. While for the most part LDWG members recognize that one can do better than legal maximums (especially the public agencies), nevertheless they are both liable for the cost of cleanup and major property owners along the river. Great structural constraints motivate them to value the riverbank as property, and to see the cleanup primarily as a process of property rehabilitation.

Those structural constraints are reinforced by the fact that most of the holdings, for both Boeing and the public agencies, are portfolio investments. They are not used by the members of LDWG for their operations but are investments that are valued in terms of what they can bring on the market. The main exception is the Port, whose holdings include the land on which port operations take place. But even the Port has extensive portfolio holdings as well. So for LDWG the overriding goal is to carry out a cost-effective cleanup that brings pollution levels down to legal requirements. That cleanup would result in the property near the river, much of which is unsaleable brownfield, to return, rehabilitated, to the market. The motivation for anything beyond is largely an optional desire, and it is always disciplined by the need to minimize costs and ensure marketable property. For the DRCC, on the other hand, the wider goal is a more comprehensive cleanup that makes the river more fully inhabitable by humans and other species in the long term. Such fuller inhabitance aims at not only a more extensive removal of pollution but also at least some measure of ecological restoration. That does not mean a complete return to the meandering river of the nineteenth century; residents of South Park and Georgetown of course do not want a river that floods their neighborhoods every few years. However, the DRCC aims to keep alive the notion that the Duwamish is a river, that it should operate in as natural a way as possible, and that it is inhabited. While the idea of a fully functioning riverine ecosystem may seem laughable in the face of the massive industrialization along the river, DRCC member IM-A-PAL foundation, in the person of activist John Beal, has proved it might be possible. Beal has restored a once mostly lifeless small tributary of the Duwamish, Hamm Creek, to fairly robust ecological health (Sato 1997). The river, Beal's restoration argues, isn't dead yet.

In claiming inhabitance in that way, the DRCC is claiming, to put it in Lefebvre's terms, a right to appropriation, a right for inhabitants to fully and completely use the watershed. To buttress that claim, they make a clever case about ownership. They insist that the river is in fact *unowned*. That is, the river is not private property. Moreover, it is not even public property in the sense that it is owned by a public agency. It is, rather, not owned at all. It is, therefore, truly a commons, a public trust, a space that is the collective responsibility of all its inhabitants. To be more specific, the *banks* of the river are owned, either by private firms or public agencies, but the river itself is not. And Superfund's mandate is specifically focused on the river itself. As a result, the DRCC claims, property rights have little standing in the cleanup; instead a regime of something more like collective stewardship should govern the river and its cleanup. They argue that whatever the actual decision-making structures that govern the cleanup, decision-making authority in the cleanup should *properly* rest with the collective inhabitants of the watershed. Any other claims to authority are usurping the legitimate authority of inhabitants. LDWG has certain obligations because they are the PRPs, but according to the logic of an unowned river, they are not the legitimate decision-makers. Therefore, public participation in the cleanup should not be an addendum to the primary decision-making process. It is rather fundamental: inhabitants have a right to decide what happens to the river they steward. To put it in Lefebvre's terms again, as a consequence of their right to appropriation, the DRCC is also claiming a right to participation. They are claiming, in other words, a right to the city.

Despite those claims, of course, the actual governance structure of the cleanup is deeply influenced by the property rights logic. But the cleanup's governance is highly complex, and it leaves extensive room for the DRCC to press their alternative claims. LDWG, as we have seen, is the lead entity. They carry out studies to evaluate pollution levels and contract specialty firms to plan and conduct the cleanup. The EPA and Ecology jointly oversee their work. CERCLA law sets out two ways Superfund sites can address pollution. The first way is to declare "early action sites" (often also called "hotspots") where there is a severe threat to public health or the environment and immediate action must be taken. The second is a longer-term remediation that seeks both to undo much of the harm pollution had caused in an ecosystem and to prevent further pollution. That latter, more comprehensive action can take years and can only be conducted in a Superfund site. In the Duwamish case, the cleanup involves both sorts of actions. Of course, the first actions have been early-action, hotspot cleanups. Five principal sites have been identified as hotspots: the Norfolk CSO, Duwamish/Diagonal way, Terminal 117, Slip 4, and Boeing Plant 2 (USEPA and Washington State Department of Ecology 2003) (see Figure 4.11). The cleanup of the first two has been partly completed, although not without incident. The second two have a completed plan and await actual cleanup, and the last is beginning the planning phase. Typically, all four members of LDWG do not participate actively in a hotspot cleanup. Rather the primary PRP takes the lead, and EPA and/or Ecology approves their actions, depending on the site.[2] The PRP can either be the direct polluter, as was King County for the CSO that was polluting the Duwamish/Diagonal way site. Or, it can be the landowner of a site that was polluted by a now-defunct firm, as is the case at Terminal 117, where the Port of Seattle owns a site that an asphalt company polluted in the past. For Slip 4, the City and County have taken a joint lead, and for Boeing Plant 2, Boeing will lead. So to an extent, for each site there is a different entity and different agents—each with different interests and different personalities—making decisions. The DRCC's role is to represent the community to EPA–Ecology as they evaluate the studies and plans LDWG produces for a given hotspot. So there is a constantly shifting decision-making terrain, depending on the site, that DRCC (and the other groups) must negotiate. That variability imposes burdens on DRCC, but it also creates opportunities. It is helpful to examine one hotspot in detail to get a clearer picture of at least one example of how the decision-making process works. While one example cannot completely capture the political complexities of the Superfund site as a whole, it can provide at least a better sense of what the politics look like.

The site at Terminal 117, also known as the Malarkey Asphalt site because of the principal polluter, is along the river in the South Park neighborhood (see Figure 4.12). Very high levels of PCBs have been found on both the property itself and in the river next to it. Moreover, PCBs have been found in the front yards of nearby houses, as well as on the road leading out of the property. The site and its surroundings pose a clear threat to both human health and the area's ecology. Malarkey is now out of business; they stopped operating on the site in 1993. The property is currently owned by the Port of Seattle, and so they are the lead agency from LDWG handling the cleanup. The cleanup has had two distinct phases. The first phase concentrated on the river and shoreline. The Port hired an engineering firm to study the pollution and draft a cleanup plan. Normally in a Superfund cleanup, the lead entity conducts studies and drafts their resulting plans, submits them to the EPA, who evaluates and approves them. Only then does the EPA seek input from the CAG. In the Duwamish case, an arrangement was worked out early in the process whereby the DRCC is granted access to the *draft* studies and plans when they are submitted to

Figure 4.12 South Park neighborhood, showing Terminal 117 site (Source for satellite image: Google Earth)

EPA–Ecology. DRCC then works with their expert (hired through the Technical Assistance Grant) to evaluate the draft proposals and formulate critiques and alternatives. Based on both the DRCC's feedback and their own evaluation, EPA–Ecology then requires revisions from LDWG. The revised document is then what EPA–Ecology approves. There is no formal requirement to grant such enhanced public input. The Duwamish arrangement is ad hoc. EPA–Ecology and LDWG understood that the DRCC was broad-based and well-organized and probably could not be contained within the more limited traditional process. The presence in DRCC of an organization that specializes in environmental lawsuits, for example, suggested they were not inclined to be docile. To an extent then, enhanced access is a way to ensure DRCC remains inside the process, rather than pursuing disruptive strategies outside it.

Early access to documents greatly increases DRCC's ability to meaningfully participate in the decision. After EPA–Ecology approves LDWG's plan, they then hold a public meeting to get feedback from citizens. In other contexts, in Superfund and beyond, such public meetings are typically impoverished: the decision-makers show a slick, well-thought-out presentation of their plans and ask the public to comment. An unorganized public made up of a variety of people interested in the issue for different reasons attempts a response. They take turns at the microphone offering a wild range of different comments, some lucid, some uninformed, some incoherent. Each speaker gets only a couple of minutes and is often cut off before they finish. Usually such meetings are more pageants to perform the idea of democratic decision-making (if poorly) than meaningful opportunities to realize it. The Duwamish case is different. Not only has the DRCC had input into the approved plan, they have also used their access to the plan to organize a community response ahead of the public meeting. They usually bring their technical expert to Seattle (he is based in Virginia) to participate in a community meeting whose goal is to develop a coherent and unified response to LDWG's plan. Those meetings are open to anyone, but generally they involve representatives from DRCC's member organizations and members of the most effected community. In the Terminal 117 case, that meant South Park residents and their association played a lead role in the meetings. At the meetings, the expert helps DRCC members and residents understand the technical elements of the plan (e.g. the pros and cons of dredging vs. suction removal of sediments). The community members help the expert and DRCC leaders understand their everyday experience with and concerns about the site. Ideally, the conversation leads to a collective shaping of a shared response to the draft plan.

At the public meeting then, LDWG's consulting firm presents the results of their study and the plan they have developed. EPA representatives offer well-worn encouragement to public feedback; "we want to hear from *you*" is a typical phrasing. But then, instead of a mishmash of different individual responses, the DRCC, because it is the Community Advisory Group, is given significant time and first opportunity to offer their response. Either their coordinator or their expert offers an equally polished, informed, and coherent presentation of the community's response to the proposed plan. After the DRCC presents, the floor is opened to other speakers. While some offer additional comments, the typical sentiment offered by these remaining speakers is, in the words of one, "the DRCC speaks for me." And the meetings have been full-to-overflowing. Speaker after speaker comes to the microphone to reiterate the DRCC's argument, an argument they helped to shape. As a result, EPA–Ecology cannot simply dismiss the public response as uninformed, contradictory, or beyond the pale. They are in fact organized, polished, scientifically sophisticated, and politically savvy. Now, while this public input is advisory only, it is so competent and it is coming from such a broad-based and well-organized "community" that EPA–Ecology takes it seriously both because they respect its content and because they suspect there could be significant political repercussions should they brush it aside.

In the Terminal 117 case, the first public meeting followed this script fairly closely. The DRCC's coordinator presented their response. She mostly offered a few critical questions (that the DRCC asked to be evaluated further) and a few alternative suggestions for the actual cleanup. Subsequent speakers supported and added to her comments. On the whole, the meeting was cordial and there was fairly widespread satisfaction with the plan. However, DRCC did make one fairly significant request: they wanted the Port to sample the upland part of the site for PCBs as well. The site is bifurcated by a small river bluff, and the original cleanup plan was to investigate and clean down at the bottom of the bluff, in the river bank and sediments (recall that Superfund's mandate is for the river specifically). DRCC wanted the Port to also sample on top of the bluff, where the asphalt plant had been. And they got their wish. EPA–Ecology approved the Port's riverbank plan, but it required the Port to sample the upland area. When they did so, they found alarming concentrations of PCBs. The upland samples had high enough concentrations that cleanup was reprioritized, and the Port set about formulating a new plan to clean the upland area first.

That upland plan was met with quite a lot less favor by the community. The PCB concentrations were physically much closer to residents' everyday routines, and the discovery of PCBs beyond the property in people's yards had made residents particularly concerned. The samples on the site found PCB concentrations as high as 9,200 parts per million, far beyond the overall legal maximum of 25 ppm (the maximum for residential use is 1 ppm) (Royale 2006; Sanga 2006). The Port's plan called for cleaning up the site to achieve levels of 25 ppm, with a zone of 10 ppm at the surface, to cap the site with asphalt, and surround it with a fence. That level would make it possible to use the site for an industrial facility, although employees would need to wear protective equipment to work there. At 25/10 ppm, it would not be safe for everyday use, which is why the fence was part of the plan. 25/10 ppm would likely also preclude other uses, such as commercial, residential, or a public park. The Port's argument, supported by EPA–Ecology, was that such cleanups are designed to return polluted land to its original use, which in this case had been an industrial facility. For the Port, an important consideration was that a cleanup to 25/10 ppm would bring the property back inside legal limits, and it could be sold on the market. Residents and the DRCC again met in advance of the meeting, and again prepared a cohesive, informed, and articulate response to the plan. Their message was clear: they wanted 1 ppm, not 25/10 ppm. That level would allow all uses at the site, including a riverfront park, which was the preferred vision that emerged from the community meeting. At the EPA's public meeting, the DRCC's technical expert gave their argument against 25/10 ppm and for 1 ppm, and then a chorus of community members came to the microphone to reiterate DRCC's message: the proposed 25/10 ppm plan did not do enough. There were two main thrusts to their argument. The first concerned the use value of the site over and above its exchange value. While a possible industrial use was good for the Port in that it would allow it to sell the land, and while it *might* benefit some South Park residents in terms of jobs, the health risks to neighborhood inhabitants were too great to ignore. Moreover, the site would be essentially closed to residents, since only those hired by the possible future firm would use the site. Far more beneficial to residents, the community argued, would be a riverfront park that all could enjoy. The second part of the argument was framed in terms of justice. South Park has a long history of bearing a disproportionate share of environmental hazards, of which PCBs in their lawns were only the most recent example. Given that history, they claimed, the Port has an obligation to do *more* than a bare-bones, minimum-required cleanup. South Park deserves better, they argued. It deserves the best the Port can do. It deserves a maximum cleanup, maximum protection from hazards, and maximum amenities. Representatives from EPA–Ecology continued to stress that Superfund law requires a cleanup to enable a return to "historical uses," that the Port's plan was in compliance with federal and state law. But South Park residents eloquently claimed that historical uses were part of a long legacy of environmental injustice, that they deserved better than the minimum.

The community's arguments were backed by a measure of political muscle. Seattle City Councilmember Richard Conlin was present at the meeting, and he submitted a letter to EPA–Ecology, signed by the entire City Council, that supported DRCC's position. South Park is in the City of Seattle, and so the residents are Council constituents.[3] Moreover, several Councilmembers have a particular interest in improving public participation, and especially in defending the "Seattle Way." Conlin is particularly keen on this point. But certainly the most important factor was that the City intends to annex the land once it is cleaned up. The Port currently owns the land, which is presently unincorporated and therefore under the jurisdiction of King County. Of course, since the Port is paying for the cleanup, the City would prefer it be as comprehensive as possible so the land they annex can be planned for any use. Despite the Council's letter, however, the Port representatives and EPA–Ecology insisted on the historic uses argument and approved the plan at 25/10 ppm.

However, that decision was not yet final. The Port Commission had to approve the plan. In Seattle, the Port's governing body is a five-person commission. Each seat is elected at large by King County voters. That elected body was much more receptive to the community's call for a 1 ppm cleanup, which South Park residents and the DRCC articulated at a Port Commission hearing on the issue. They were again backed by the Seattle City Council, who renewed their request for 1 ppm. Also supporting the community was Zach Hudgins, Washington State Representative for the 11th District, in which Terminal 117 sits. And among some residents the claims took a somewhat more threatening cast. One promised the commission that South Park residents "will not live quietly with hazardous waste" and intimated that lawsuits were on the horizon (Scott 2006). The Port commissioners decided to reject their own staff's plan and instead require a cleanup to 1 ppm. One commissioner, quoted in the local press, felt the Port needed to rebuild its reputation and "to earn back the trust of the community" (Modie 2006). However, the Port also called on the City to help pay the added cost for 1 ppm, since they stand to benefit directly from a more thorough cleanup. The Commission's 1 ppm plan is where the matter currently stands. EPA–Ecology has promised to try to meet that standard, and new glitches could arise at any time. When DRCC's director was asked if this was a total victory for the community, she said cautiously, "it seems that way" (Modie 2006). That response is characteristic. She is frequently skeptical when evaluating any apparent victory. She understands the ongoing nature of the struggle, and the possibility that any decision can be reversed. In this case, even if the 1 ppm plan goes forward, the site will be the subject of future struggle, both in debates about the more comprehensive, riverwide cleanup, and in zoning, land-use, and development struggles over the property once it is cleaned.

The case of Terminal 117, and that of the Duwamish cleanup more generally, suggests several specific lessons about democratic movements and neo-liberalization. The first is that *it is possible to form a stable network of equivalence around the agenda of inhabitance*, and that network can have real (if modest) success in resisting neoliberalization. The DRCC is a broad coalition, and its members have divergent interests. But they have constructed a sufficiently equivalent way of understanding the river and its basin, one that allows them to act in concert and make a coherent set of claims on LDWG and EPA–Ecology. The breadth and diversity of their coalition is crucial to their success because it solidifies their ideological claim to represent "the community." The EPA requires explicitly that CAGs represent the community as comprehensively as possible, and they take CAGs more seriously when they are diverse and broad. But just as critical is how coordinated the coalition's message and actions are. If they could not articulate a coherent vision, and if they could not make their claims effectively and in concert, their influence would be greatly diminished. So they must be both broad and coherent; they must achieve equivalence by being simultaneously the same and different. It appears, at least so far, that the DRCC is able to rise to that challenge.

The second lesson is that despite appearances, *under neoliberal hegemony there are real opportunities for democratic resistance*. While neoliberal governance structures diminish democratic decision-making in important ways, that process is by no means total. Many old structures for democratic participation remain open, and neoliberal governance arrangements frequently open up new opportunities for significant democratization. Under neoliberalization governance becomes increasingly ad hoc, and therefore indeterminate. LDWG, for example, is new; it has never done this kind of thing before. It is learning as it goes, governing by trial and error. It doesn't have a routinized way to deal with a community movement like the DRCC. That indeterminacy produces a very open system with relatively unstable (or, relatively more fluid) elite alignments, two key elements of political opportunity structures that social movements can take advantage of (McAdam 1996; Miller 2000). However, both old and new opportunities are only latent; to be realized they must be actively exploited. Mobilized networks of equivalence must identify and realize their opportunities. Moreover, they must struggle to expand nascent opportunities and invent new ones. The governance structure on the Duwamish presents all sorts of democratic opportunities that the DRCC has searched out, exploited, and expanded. They explicitly hope their efforts not only help the public participate more fully in this cleanup, but also set a precedent that future Community Advisory Groups can build on. More generally, the democratic innovation on the Duwamish could be applied creatively to a bundle of urban political issues where federal requirements for public participation exist. Anything subject to the jurisdiction of, for example, ISTEA (transportation), HOPE VI (housing), EPA (Brownfields, pollution, environmental justice, etc.), and ESA (endangered species), have similar provisions that could open the door to similar kinds of mobilization. Subverting the neoliberal practice of "fast policy transfer," we might imagine a "fast resistance transfer" whereby democratic innovation is spread through networks of movements (Peck and Tickell 2002b). Ideas from other places could not, of course, be applied without modification to the new context, but neither would each movement need to construct anew their democratic resistance to neoliberalization. That kind of networked, collective, political learning can spread quickly, and it can offer all sorts of concrete advantages to social movements struggling to maintain themselves under neoliberal hegemony.

A third lesson, which flows from the second, is that in a climate of relatively open governance structures, *movements must be more nuanced in assessing the age-old dilemma between playing the game and remaining outside it* (Young 2001). Recall that the EPA and LDWG hope the enhanced participation arrangement will keep the DRCC inside the process, rather than agitating outside it. The DRCC is aware that they are choosing to play the game, but they are also actively working to remake the game as they go along. The ad hoc enhanced participation process is itself evidence of that. After they decided to apply to play the game by becoming the CAG, the DRCC then immediately pressed to remake the game by expanding the influence a CAG has. While to an extent they are being "channelized" into the structures set up by regulatory agencies, they are always working to question those channels and move beyond them, to reimagine the assumed limits of public participation. In other words, they reject the choice between (1) passively playing the game the EPA and LDWG want them to play and (2) refusing the game and challenging it from the outside. Rather, they perceive that the openness of these particular governance structures offers a third option to play the game in order to remake it. Their advice to other movements might be: build momentum in the channels you are offered and then overspill them every chance you get. To date, that has been the DRCC's approach. They retain a sense of themselves that Mouffe might call agonistic: they participate *in* the process, but they are not *of* the process. For the most part, they see LDWG and EPA–Ecology not as partners but, to use Mouffe's term again, as adversaries, as groups with political agendas that conflict with their own. That distinction is important because it means the DRCC remains always aware of the possibility of leaving the game. They are engaging the process not to discover outcomes in the common interest of all participants; rather they are engaging it strategically, to get the most they can for the agenda of inhabitance. If they believe their play-to-rework strategy is no longer the best option for that agenda, they retain the option to, in the words of their director, "picket on the banks."

The fourth lesson suggested by the Duwamish case is that *inhabitance can be a viable basis around which to construct equivalence.* I should be clear that the DRCC does not articulate what they share precisely in that way. They do not talk about an abstract idea of "inhabitance" and consciously articulate "equivalence" on that basis. At the same time, they do consciously construct a shared sense of the river, and they have a clear sense of themselves as wanting something distinctly different than what LDWG wants. In each cleanup action to date, the members of the coalition have agreed by consensus on a joint way to proceed that furthers first and foremost the interests of those who inhabit. While they recognize the interests of property owners, they see them as at best a secondary consideration.

A last lesson is also important. It suggests that *there is often a real strategic need for cooperative, deliberative politics* inside *networks of equivalence.* I have argued in general for democratic attitudes that favor agonistic social mobilization over deliberative cooperation. However, the Duwamish case illustrates the importance of understanding the nuance that Nancy Fraser's (1990) "subaltern counterpublics" raise. Inside a subaltern counterpublic, there is often a strategic need to work toward consensus so that the counterpublic can develop an oppositional understanding and agenda. That allows them to function more effectively as "spaces of withdrawal and regroupment," as Fraser puts it. In confronting the wider public through agonistic struggle, it is usually enormously more effective for counterpublics to present a coherent front. In their community meetings in advance of public meetings, the DRCC does use deliberative methods to work together toward a consensus that can be presented as the response of "the community." It also allows them to nurture and solidify their sense of an equivalent agenda. However, any critique of deliberative democracy would of course apply here: any consensus, even within the counterpublic, must exclude some and include others, and it will favor some inside the group over others. In the case of DRCC, an identifiable excluded group has been Spanish speakers in South Park. Despite concerted efforts (they have recently hired a native Spanish speaker to serve as Bilingual Outreach Coordinator), the DRCC have so far been unsuccessful in creating meaningful opportunities to include the voices of the ESL population in their deliberations. That lesson is important because it represents the tension that all networks of equivalence must face. They must find a way to link together the irreducibly different agendas of their network, enough to present a coherent face and message to the wider public. Equivalence is a less reductive way to do that than "one big union," but any process in which diverse groups come together to act in concert will necessitate some process of exclusion, favoring some interests over others. Each network must weigh such problems of internal exclusion and inequality against the benefits of acting together to make claims on the wider society. Such tensions are inherent in the process of building equivalence, and it is an ongoing politics. The precise nature of equivalence must be continually revisited, because each group must continually reaffirm their connection to the network, deciding that the power that comes from acting together outweighs the exclusions inevitably produced by a politics of agreement.

In the Duwamish case a movement comprised of both inhabitants and non-inhabitants has come together to claim a right to inhabitance. The next case, that of homeowners in Los Angeles, demonstrates that movements of inhabitants are not always progressive. In Los Angeles, inhabitants did not make clear claims to inhabitance, at least in the way Chapter 3 understands that concept. They did not build networks for inhabitance, nor did they resist neoliberalization. The case thus raises important red flags in our thinking about movements of inhabitants. It helps us understand the significance of the distinction between movements of inhabit*ants* and claiming inhabit*ance*. It allows us, in short, to better grasp what is at stake in claiming a right to the city.

LOS ANGELES IN BRIEF

The Los Angeles case differs from the Seattle cases in several ways. The first difference is temporal. The Seattle cases are all contemporary, but the Los Angeles case recounts politics that took place primarily in the 1990s. The evidence presented in the narrative was gathered during fifteen months of embedded field work in the city. The goal of the project was to understand the role that homeowners associations played in Los Angeles politics. It employed interview, observation, and archival methods to obtain data, and grounded-theory techniques to analyze it. As with the Seattle case studies, where no specific source is given, information was taken from field notes. In the 1990s, politics in Los Angeles were both similar to and different from those in contemporary Seattle. On the one hand, Los Angeles was very much integrated into global circuits of capital, even more so than is Seattle today. In was a global center of aerospace, high technology, entertainment, shipping, and biomedical activity. Like other global cities, it was home to corporate headquarters and other command and control functions for the global economy. It was a primary destination for immigration to the United States from Latin America and Asia. It embodied all the characteristics of a first-order global city. However, unlike Seattle, that key role in the global economy was not overlain by any particular commitment to innovative democratic politics. The political traditions in Los Angeles were much closer to the machine politics of Eastern cities than to the progressive traditions so common for cities in the western United States. While some progressive reform took place in the early part of the twentieth century, the contemporary governance structure of the City of Los Angeles was a ward configuration in which each of fifteen councilmembers represented a council district of about a quarter of a million people. To a significant extent, those wards, large as they were, had distinct ethnic profiles: the first was predominantly Latino, the 11th was predominantly white, and the 10th was historically mostly African-American, although that profile was changing.

That ethnic politics was a part of a wider demographic shift in the city. Throughout the twentieth century Anglos made up a distinct majority of the region's population. Starting in about the 1970s, though, the Latino population began to increase, and by the 1990s Anglos had become a minority. While the formal political power of Anglos had persisted past the demographic tipping point, that power has recently begun to wane. Latinos are now dominant demographically, and they are steadily becoming dominant in elected office as well (Davis 2001). But such was not yet the case in the San Fernando Valley, part of the City of Los Angeles and the location of all the homeowners associations that were the focus of the research. The Valley was still predominantly Anglo. While that too was beginning to change, politics in the Valley were still dominated by Anglo interests. Thus politics in Los Angeles, far more than in Seattle, involved an explicit contestation between an old-guard Anglo middle- and upper-class, and a rising Latino population, most of whom were lower-to-middle-class, as well as its growing vanguard of elected politicians. Added to that mix were smaller but significant populations of African-Americans and Asian-Americans. In city politics, the Balkanized city council is the dominant agent, because one legacy the progressive movement did leave was a weak mayor structure that left most control to the council.

In that pitched-ethnic-battle and machine-style political context, innovative deliberative practices for public participation were not nurtured. In Seattle, where the Anglo middle-class remains demographically unthreatened, the power structure has the luxury of allowing some of the kind of cooperative, collaborative political ethic that deliberative democracy demands. In Los Angeles, such cooperation and civic high-mindedness appeared quite a lot more fanciful than in Seattle. Politics were at least agonistic, and more often antagonistic. Collaboration in pursuit of the public good was not something embedded in the city's political culture. Urban politics in Los Angeles were a struggle among competing elites, some elected, some not, who amassed influence, favors, and finances to pursue their particular agenda (Hunter 1953). What public participation there was used more traditional means: public hearings, voting, and social mobilization. There were few serious structures in which the public could meaningfully influence decision-making. If they wanted to be heard, they needed to organize and mobilize. As a result, not surprisingly, the city was a center of innovative social movements, perhaps precisely because it almost entirely lacked innovative deliberative-democratic experiments. It is home to widely-known movements such as Justice for Janitors, the Labor/Community Strategy Center (which spawned the Bus Riders Union), Mothers of East Los Angeles, and the Los Angeles Alliance for a New Economy, among many others. On the other side of the political spectrum, but no less influential, was a movement among affluent homeowners in the San Fernando Valley first documented by Mike Davis (1990) in his book *City of Quartz*.

Homeowners in Los Angeles

The case of homeowners in Los Angeles illustrates the difference between movements of inhabit*ants* and movements for inhabit*ance*. That is, movements of inhabitants can have all sorts of agendas that may or may not be consistent with democracy or with the right to the city's agenda for inhabitance. As I have presented it, the right to the city entails an agenda for urban space that values inhabitance and the use of space over and above the neoliberal agenda of property rights and market exchange. The case of homeowners in Los Angeles presents an example of a movement of inhabitants whose agenda, while it has some points of connection to inhabitance, generally diverges from the essence of the right to the city.

The San Fernando Valley (see Figure 4.13) was annexed to the City of Los Angeles in 1915 as part of a deal that brought water south from central California to the semi-arid city. By itself, the Valley is larger than most U.S. cities, at almost 1.7 million people in the 2000 census. While it remains predominantly Anglo, its Latino population is growing larger. And there is a distinct geography to that Latinization. The relatively affluent Anglo population tends to live in a ring around the Valley, on the slopes of the mountains that contain it. The growing Latino population tends to live on the wide Valley floor, and it is moving westward across that floor, from its historic center in the East Valley around the City of San Fernando, site of the old Spanish mission in the area. The homeowners movement I examine is based in the South Valley, in the neighborhoods (from west to east) of Woodland Hills, Tarzana, Encino, Sherman Oaks, and Studio

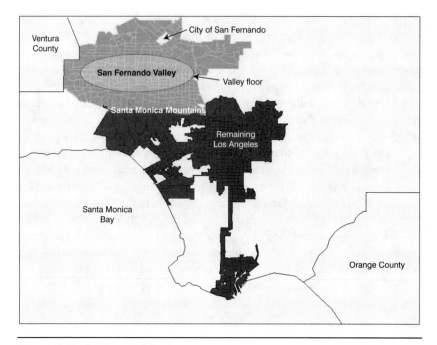

Figure 4.13 The San Fernando Valley in context

City. That area remained relatively Anglo and relatively affluent according to the 1990 census (see Figures 4.14 and 4.15). While homeowners' associations have been the focus of much scholarly attention in recent years (e.g. McKenzie 1994, 1998; Blakely and Snyder 1997; O'Neill 1992), most of that work has focused on mandatory associations that people join as part of moving into an (often gated) common-interest development, the increasingly common single-family-neighborhood development model that works much like a condominium, with public areas owned in common among the residents. The Valley associations were very different. They were voluntary organizations in older residential neighborhoods. They were run by volunteer staff. They were very large and very politically active. They constituted a social movement among people who were relatively affluent, Anglo, and professional compared with the population of both the Valley and the city as a whole. Through their committed activism, they were able to influence decision-making on both the neighborhood and citywide scales.

Taken as a whole, their agenda was to preserve a particular kind of idealized geography. They wanted to save the last remaining vestiges of a Valley that they remember to have existed in the past, one that typified the suburban ideal: low-density, single-family houses, and respectable clean, verdant landscapes (Purcell 2001b). It is a vision that Robert Fishman (1987) has labeled "bourgeois utopias"—places that embody the ideals of private ownership, nuclear families,

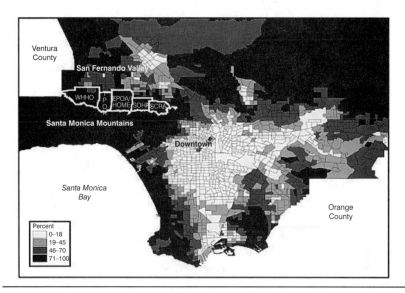

Figure 4.14 Homeowners associations with percentage white in tract. Associations mapped are: Woodland Hills Homeowners Organization (WHHO), Tarzana Property Owners Association (TPOA), Encino Property Owners Association (EPOA), Homeowners of Encino (HOME), Sherman Oaks Homeowners Association (SOHA), and Studio City Residents Association (SCRA) (Source for data: 1990 U.S. Census)

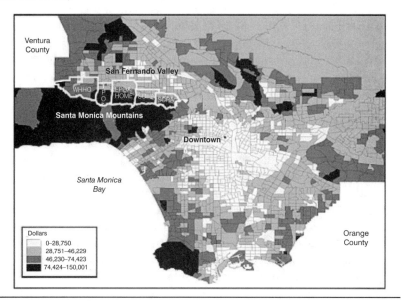

Figure 4.15 Homeowners associations with median income in tract. Associations mapped are: Woodland Hills Homeowners Organization (WHHO), Tarzana Property Owners Association (TPOA), Encino Property Owners Association (EPOA), Homeowners of Encino (HOME), Sherman Oaks Homeowners Association (SOHA), and Studio City Residents Association (SCRA) (Source for data: 1990 U.S. Census)

harmony with nature, exclusion, segregation, and privilege. Robert Fogelson (1967, pp. 144–5) argues that imagination is especially endemic to Los Angeles, a city shaped greatly by the dreams of Midwestern Anglos who

> came to Los Angeles with a conception of the good community which was embodied in single-family houses, located on large lots, surrounded by landscaped lawns, and isolated from business activities. Not for them multi-family dwellings, confined to narrow plots, separated by cluttered streets, and interspersed with commerce and industry. Their vision was epitomized by the residential suburb—spacious, affluent, clean, decent, permanent, predictable, and homogeneous—and violated by the great city—congested, impoverished, filthy, immoral, transient, uncertain, and heterogeneous.

Homeowners imagined this ideal to have been achieved in the Valley's postwar past, and they felt it under threat in 1990s Los Angeles, undermined by a host of dangers, from rampant development, increasing density, growing numbers of immigrants, worsening traffic congestion, escalating crime, distant and aloof city government, deteriorating public services, and substandard schools. Their agenda was to stop these threats and defend their idealized suburban past.

The movement was, therefore, very much a movement of inhabitants. However, it was not a movement for inhabitance in the way Lefebvre conceives it. It was not concerned to defend urban space-as-inhabited against urban space-as-owned. Rather it sought to preserve for *some* inhabitants a certain kind of urban space that they valued, and prevent the spread of other kinds of urban space—and other kinds of inhabitants—that they disapproved of and feared. For example, during the research period homeowners worked hard for enforcement of a ban on garage apartments. Some homeowners in the area had converted their garages into substandard apartments that were often occupied by recent immigrants. That practice resulted in greater density and greater ethnic diversity in the homeowners' neighborhoods, as Latinos (and East Asians, to a lesser extent) were the predominant inhabitants of these dwellings. The homeowners associations vigorously opposed the illegal conversions. But they were not advocating for the inhabitants of the garages, to bring the dwellings up to code. Rather they wanted the practice to end, and by extension they wanted the recent arrivals to no longer inhabit the area. So the homeowners' movement was primarily a conservative one, concerned to preserve their existing privilege by excluding new development and new inhabitants.

The movement was, it should be said, not just a movement of inhabitants but also a movement of property owners. Only one association, the Studio City Residents Association, included renters among its members, and even they were made up almost entirely of homeowners. So activists were simultaneously inhabitants and property owners. I have argued that the importance of property values to such movements is usually very much overstated (Purcell 2001b). Homeowners movements are in fact mostly concerned about quality-of-life issues that impact their daily routines. Nevertheless we should not entirely discount a concern for property values among activists. The resale value of one's house, in which a significant portion of one's assets are usually invested, is important to people (Oliver and Shapiro 1997). To the extent the Valley was changing against homeowners' wishes, and they were thinking about selling and getting out, the value of their property was a concern to them. But it is most accurate to say that their activism was mostly that of inhabitants concerned about the *character* of their neighborhood. They intended to stay, not sell, and they wanted to defend and recapture the kind of place they thought their neighborhood should be.

In practice, that agenda commonly led them to oppose property developers, owners whose interest was to maximize the value of their holdings. The homeowners were concerned with their everyday use of urban space: traffic congestion, parking availability, garish signage, noise, pollution, etc. However, their agenda, again, was to preserve a particular kind of landscape for a particular subset of inhabitants. While there are certainly points of contact between that agenda and that of inhabitance, those were mostly contingent: the movement's primary goal was not to promote inhabitance-in-general against ownership-in-general. They did not claim for all inhabitants a right to inhabit space, but rather a right for affluent Anglo inhabitants to *their* notion of what the city should be. For them, recent-immigrant inhabitants were just as much a threat as the Jiffy-Lube developer on the corner. The Los Angeles case illustrates that it is not enough to have a movement of inhabitants. Resisting neoliberalization requires also that the movement pursue, either explicitly or implicitly, an agenda *for inhabitance*. Otherwise, as happened in the Valley, the movement can take on a reactionary, exclusionary character that may or may not resist neoliberalization, and certainly stands in the way of a more just and civilized city.

To see more fully why the movement was not claiming a right to the city in the way I have characterized it, we must look more closely at the specific politics they pursued. Their everyday concerns were about development, traffic, congestion, graffiti, and landscapes in their neighborhood (more detailed data from the research is reported in Purcell 2001b). They closely monitored new developments and lobbied both the developers and their councilmember for more parking and vegetation, better traffic mitigation, reduced density, and the preservation of viewsheds and open space. Beyond these more local concerns, they also engaged in citywide efforts to restructure politics so that they might participate more effectively. They were involved, for example, in efforts to get more favorable terms for the public in building a downtown basketball arena, to rewrite the City Charter, and to help the Valley secede from the City of Los Angeles.

The effort to secede was the most spectacular of those campaigns. The large south-Valley homeowners associations allied with small-business interests, realtors, and small-firm land development interests to help the Valley secede from the City of Los Angeles and become an independent municipal government (Purcell 2001a). Together they formed an organization called Valley VOTE (Voters Organized Together for Empowerment). The coalition was quite large, and it was diverse in the sense that it brought together groups concerned with exchange value and business climate with groups concerned with use value and neighborhood character. Those two overarching factions, each made up of many organizations, are of course habitual foes in the politics of land development. However, they allied around an equivalent interest in Valley secession. They therefore constituted a network of equivalence whose goal was to boost the Valley's formal political power, power that was diluted because the Valley was part of the larger city government.

As a result, they argued for more "local control" for the Valley, which they claimed would be a more democratic political arrangement. In making that claim, they evoked a Jeffersonian notion of agrarian democracy, where minimal and decentralized government allowed people more direct control over their lives. While we might question the movement's commitment to democracy, it clearly wanted to increase local control. As the South Lake Union case suggests, localization is not the same thing as democratization (Purcell 2006; Brown and Purcell 2005; Low 2004). Any localization of power begs the question of what one wants to *do* with that power. Valley business interests were classic neoliberals; they wanted primarily to improve the business climate in the Valley. They felt local control would allow them to better reduce regulations, taxes, and bureaucratic hurdles for businesses. Current city government, they felt, was too large, inefficient, and intrusive; it was squelching businesses and preventing new investment in the Valley. Homeowners groups also felt the City was too large, but not in the same laissez-faire, anti-big-government way the business groups did. Rather, they felt the government was too large to be responsive to its citizens, and more importantly, too large for their organizations to have much influence over its decisions. Although their organizations gave them some influence over their councilmember, each organization was only a small part of any one council district (see Figure 4.16). They were not always their councilmember's top priority. Moreover, the other, non-Valley council districts were predominantly non-white, and they had very different agendas that were at least not overly concerned with—if not directly hostile to—the needs of the Valley. Homeowners hoped that in a smaller Valley City, with council districts that more closely matched their neighborhood boundaries, they could have significantly more influence over what happened in their neighborhood. They

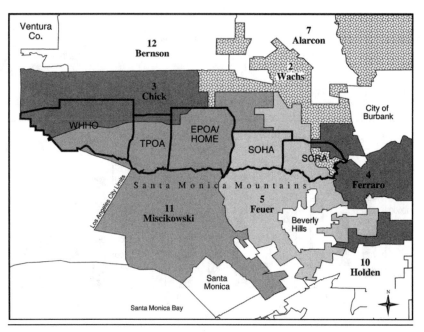

Figure 4.16 Homeowners associations with City Council districts, 1990s

intended to use such influence to pursue their agenda: preserving for relatively privileged inhabitants a particular kind of urban space that meets their needs.

In addition, while the homeowners were not greatly concerned with taxes on businesses, they were concerned with issues of collective consumption. They argued the Valley didn't get its "fair share": the tax revenue the Valley paid to the City was greater than the value of public services the Valley received in return. Their argument was in fact deeply regressive. Since the Valley is relatively wealthier and relatively less in need than the remaining City, one would expect it to pay more than it receives back (see Figure 4.17). The "fair-share" argument, couching any subsidy as "unfair," was a clever rhetorical way to oppose the redistribution of resources to poorer areas of the city. Business interests in Valley VOTE were moderately concerned with the fair share question (especially realtors concerned with improving Valley public services like police, fire, libraries, etc.), and they also understood the way that rhetoric resonated with Valley voters, who would ultimately have to approve any secession. But their concern with taxes was primarily one of laissez-faire.

So Valley VOTE was in many ways a network of equivalence in which diverse interests acted in concert to pursue a shared agenda. The network involved inhabitants, and, at least rhetorically, they called for greater democracy in urban politics. However, the network's equivalence was distinctly *not* constructed around inhabitance. Instead, it was constructed as a

	Population	Median household income, 1989 ($)	Non-Hispanic white population (%)	Median house value ($)
Valley alone	1,219,079	45,442	56.6	270,525
Non-valley city	2,259,841	30,279	26.8	226,864
Entire city	3,478,920	35,531	37.2	241,794
Entire county	8,863,164	38,016	40.8	235,960

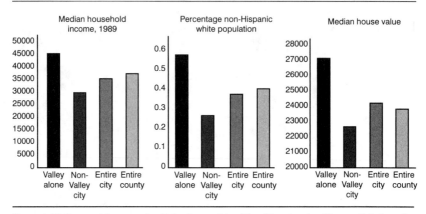

Figure 4.17 Census data comparing Valley to remaining City of Los Angeles (Source: Data from the 1990 U.S. Census)

shared desire to increase the Valley's formal political power. Business interests were clearly not concerned to defend inhabitance and resist property rights. And while Valley VOTE did hope that secession would empower inhabitants, they wanted that empowerment only for a particular subset of inhabitants, and a particular kind of inhabitance. They were not concerned, as we have seen, to defend inhabitance-in-general against property rights. Their call for democracy was not a desire for a radical democratization of decision making, but for a localization of power that would increase their influence over the decisions that concerned them.

In the end, ValleyVOTE failed to achieve secession. A citywide vote in 2002 rejected it. But in order to even bring about that vote, the coalition had to achieve significant victories, both at the local scale and in the California legislature. They got much farther than anyone thought they would. So in one sense Valley VOTE is a cautionary tale; it reminds us that mobilizations of inhabitants are not necessarily movements for the right to the city or for radical democracy. It also reminds us again of the potential for neoliberalism to partner with democracy. By pairing local control and democracy, and then allying that agenda with resistance to big-government interference, Valley VOTE was tapping into the anti-Keynesian, classical liberal strain of neoliberal thought that says that all collective (rather than individualized) solutions to problems, especially in the form of the state, are antithetical to freedom and democracy. On the other hand, despite these cautions, in another sense the case of Valley VOTE can help us move forward. It reiterates the power that networks of equivalence can produce. They are a potent political strategy, and there is every reason to believe they have the potential to subvert the hegemony of neoliberalization. But it is worth remembering that the agenda of such movements is subject to political struggle. Even when it involves inhabitants, the movement may or may not resist neoliberalization, and it may or may not aim at a more democratic city.

Another example of the activism that LA homeowners engaged in offers more hope. It concerns the development of a downtown sports arena, and it suggests that even if they do not set out to do so, movements of conservative, even reactionary inhabitants can generate the potential for anti-neoliberal resistance. In the arena case, homeowners became involved in a movement to minimize public assistance to the developers of a professional basketball and hockey arena in downtown Los Angeles. In the 1990s the Los Angeles Lakers and Los Angeles Kings were playing in the outdated Great Western Forum in the City of Inglewood and wanted to build a new arena. They entered into negotiations with city government to begin a project in downtown Los Angeles, just north of the city's convention center (see Figure 4.18). At the outset the city was represented in negotiations by Steve Soboroff, a local commercial real estate broker who was an unofficial advisor to Mayor Richard Riordan but who held no official government position. The early negotiations took place out of public view until a local reporter discovered the story and began writing columns critical of the clandestine nature of the process (Boyarsky 1997a, 1997b). The columns stirred up further opposition, primarily from populist councilman Joel Wachs, whose second district was located in the East Valley. Wachs was joined by, among other activists, Valley homeowners. Soboroff had been negotiating a deal that, as in many other cities, would be immensely favorable to arena developers (Keating 1997). Fearing the teams could find a better deal in another nearby city, he offered tax breaks, nominal rent, and at least $70 million in public money toward the arena's estimated $250 million cost (Purcell 2002b). Soboroff and most in city government felt that in order to attract an arena they needed to contribute significant public money to mitigate the cost of development. It was a classic arrangement in the era of neoliberalization. It was, moreover, fairly typical for contemporary pro-sports arena deals (Eisinger 2000).

But Wachs saw an opportunity. He understood that in the midst of the "fair share" rhetoric of secession his Valley constituents hated city government spending on anything that did not benefit the Valley directly. He denounced both the secrecy of the negotiations, and the amount of public money flowing into the deal. He formed a group called "Citizens Against Secret Handouts"

Figure 4.18 Staples Center and Downtown Los Angeles (Source for satellite image: Google Earth)

(CASH) and began sending out mailings proclaiming that people have grave concerns about "giving their hard earned tax dollars to billionaire team owners and mega-millionaire players" (Purcell 2002b). He demanded that the city drive a harder, less costly bargain. However, Wachs was only one vote on a council of fifteen, and he was the only council member willing to take a stand on the issue. His only leverage was his potential public support. CASH was thrown together quickly; there was no time to do proper organizing. But the existing network of homeowner activists was already in place. That network gave Wachs' resistance credibility. Moreover, he had real political options: in California citizen initiatives are commonly used to make policy. Wachs launched a city ballot initiative that would have rewritten city law to require that any new sports facility receiving public funds would have to be approved by voter referendum. He began campaigning for his ballot measure, speaking with great flair at neighborhood association meetings around the city and especially in the Valley. He was playing to extremely favorable responses, and it appeared as if a genuine movement was developing to pass the initiative.

Soboroff and the developers took Wachs' initiative to be a real threat. For them it would mean unacceptable delays for the process. They began asking if they could share the stage with Wachs at the association meetings, where they offered standard neoliberal economic development arguments. Investment in the city is good for all, they said; it creates jobs and boosts city revenue. If the City doesn't offer favorable terms, they claimed, the developers will build in a city that will. Residents responded to these arguments with outright contempt. At several meetings, the developers were nearly jeered off the dais. Activists saw the arena deal as yet another case of their government arrogantly ignoring their interests. As the anti-arena movement developed, it appeared Wachs' initiative might significantly delay or even derail the project. The developers began to renegotiate the terms of the deal. Eventually Wachs and the activists won: an agreement was reached whereby virtually all of the risk and cost of the development was assumed by the developers and very little by the City. In effect, the developers were welcome to build in Los Angeles, but they would receive relatively little financial assistance from the City to do so. The developers chose to absorb the costs and go ahead with the project rather than go to another municipality. Wachs' gamble paid off in a surprising political victory whereby the city took seriously the discontent among some of its citizens, and rejected the imperatives of economic competitiveness.

In this case, Wachs and his homeowner base were a relatively conservative political force. They did not explicitly pursue an agenda of inhabitance over ownership. They were not, for example, much concerned with how the development might displace low-income residents downtown. They did not explicitly claim inhabitants should benefit from public investment rather than capital. However, there was a kernel of promise here. The movement clearly opposed spending public money (*their* money, as they said) to spur economic development. It favored collective consumption spending for inhabitants instead. The outcome destabilized the neoliberal claim that cities *must* invest significant public money to attract development, a claim the movement questioned all along. The rank-and-file of the movement also roundly rejected the neoliberal refrain that economic development eventually benefits all. They felt that the developers and "billionaire team owners" were the real beneficiaries. Soboroff and his team seemed to offer their trickle-down argument almost out of habit, and they were surprised and confused when the audience scoffed.

Moreover, the movement was explicitly concerned with transparency in governance. They called into question the secretive negotiations that were conducted outside normal decision-making structures. They therefore raised an essentially liberal-democratic critique against the tendency toward unaccountable governance under neoliberalization. Lastly, to reiterate, the rank-and-file of the movement *were* inhabitants. They did not articulate their vision that way, mostly because they were fundamentally exclusive in the way they imagined their movement. They were only concerned that public spending for the arena would limit the kind of services they would see in *their* neighborhoods. They were not concerned about how it would impact other inhabitants. The movement was therefore made up of inhabitants, but it excluded most inhabitants. Nevertheless, there is potential here. It is easy to see how such a movement could be imagined inclusively, how this faction of inhabitants could ally with other inhabitants to claim the right of inhabitants to adequate public services. They could claim that their needs as inhabitants come first, before the needs of developers and property owners. As in the South Lake Union case in Seattle, such a movement did not develop in Los Angeles. But the political sentiment exists. Inhabitants across Los Angeles and Seattle respond to the idea that their needs as inhabitants (streets, housing, parks, safety, health, transportation, etc.) should be a higher priority for their government than the needs of capital. Homeowners in Los Angeles were exclusive of others because of contextual tensions having to do largely with their fears of ethnic succession. If they could work through that fear, perhaps by seeing the extraordinary political potential of a citywide alliance of inhabitants, they could begin to make claims that seriously challenged the hegemony of neoliberal assumptions. Such a citywide alliance, bridging as it would a vast collection of social differences, is admittedly a great challenge. But even in Los Angeles it is not at all impossible. As a whole, the Los Angeles case suggests some of the dangers of inhabitant mobilizations; but it hints at some of their enduring promise as well.

Conclusion: You Can Hear Her Breathing

You built your tower strong and tall
Can't you see
It's got to fall
Someday?

—Townes Van Zandt, Tower Song (1971)

In October 2005, George W. Bush spoke to the National Endowment for Democracy. Standing in front of a phalanx of thirteen American flags and a banner that read "supporting freedom around the world," the President made an ardent argument for democracy. He was introduced as someone who supports "democratic movements around the world" and is "personally passionate" about the subject of democracy. Bush's speech was an argument to remain steadfast in the "war on terror." He said that the war presents a fundamental opposition between democracy (us) and "terror" (them). Democracy stands against terror, totalitarianism, fascism, tyranny, and slavery. Democracy will win the fight, he believes, because it offers *freedom*, and freedom is what people want. According to Bush's narrative, democracy provides political freedom, and market liberalization provides economic freedom. Those two processes are paired under the common cause of "supporting freedom around the world." That twinning, which we met before, is not new. The National Endowment for Democracy was initiated by Ronald Reagan at the height of his Cold War crusade. The idea was to help reiterate the argument that the Cold War was a struggle between democracy–freedom–capitalism on the one hand, and totalitarianism–tyranny–communism on the other. As we saw in Bush's

National Security Strategy (Office of the President 2002), the democracy–freedom–capitalism package endures. Free markets and free elections are part of the same bundle, one that all "free peoples" must have. The only alternative to that package, neoliberals would have us believe, is tyranny, slavery, and suffering.

Democracy is too powerful a banner to abandon to neoliberals like Bush. Ceding that ground and sticking instead to other political concepts, like justice or citizenship, would hand neoliberals a critical tool of political legitimacy. I do not mean to say that those other concepts are not also important. Rather, if we want a better urban future than neoliberalization offers, we cannot allow it to equate capitalism and democracy. Neoliberalization produces instability by its normal operation, and so it can only survive if it can ensure its own political legitimacy. Democracy is the most powerful means to achieve that legitimacy. If democracy and neoliberalism are allowed to stand as a pair, then neoliberalism will be virtually unassailable. But more than that, if the democracy that Bush presented in his speech is to be the dominant understanding of democracy, then the promise of democracy is eviscerated. Democracy would be reduced to its very basest, most vapid formulation. Democracy has far more potential than that. Conceived of more fully, it aims at much more equal and freer polity than current forms of liberal democracy can imagine. And that is one of the many things that democratic movements can do: they can say out loud and in the streets that Bush's democracy is not the only or best form of democracy. There are other ways to think about democracy, ways that take seriously its radical claim to equality and freedom. If we don't mobilize to articulate them, defend them, and work toward them, then we will be stuck with the minimal, formal democracy favored by Bush. Perhaps worse still, we will be trapped in the neoliberal city, where competition dictates policy, ownership confers legitimate authority, and the city is valued for what it can bring on the market, rather than as a home to its inhabitants.

THE CHALLENGES

The fact that we must reclaim the banner of democracy does not mean that it will be easy. As I have tried to make clear, the neoliberalism–democracy twin is the hegemonic understanding of political-economic relations in the world today. If most people understand that we have not reached "the end of history," nevertheless Fukuyama's underlying argument, that the most evolved kind of society is a liberal-democratic and capitalist one, has become the dominant view. It has reached the level of political-economic common sense. So any movement for democratization and against neoliberalization must first undermine that hegemonic assumption. It must relentlessly document and publicize the deeply undemocratic results of neoliberalization. It must narrate the innumerable cases like South Lake Union, where the pressures of competitiveness undermine an open and inclusive decision-making process. And it must continually celebrate the radical potential of the democratic idea, a potential liberal democracy cannot fulfill.

Making the challenge of that hegemony even greater, neoliberals work actively to co-opt and absorb new democratic initiatives. They seek decision-making processes that can confer political legitimacy but that do not undermine the material interests of capital and property owners. That absorption strategy is driven by neoliberalization's continuing and deepening need for legitimacy, and so it will always be a part of the neoliberal project. Any new initiatives for democratization that successfully gain currency should expect to attract the active interest of neoliberals. If they do not explicitly stand against neoliberalization, against marketization, individualization, and privatization, if they do not reject the prioritization of property rights as the basis for land use, and if they find business interests readily agreeing to participate actively in their process, then, at least in the long term, new democratic projects are as likely to reinforce neoliberalization as to oppose it. The case of the waterfront in Seattle suggests how such initiatives can confer legitimacy but offer little challenge to dominant political-economic relations.

I have argued that democratic movements, organized into networks of equivalence, have the potential to undermine neoliberal hegemony and help create a more radically democratic urban polity. However, such movements themselves pose a third challenge: it is not at all easy to forge equivalence. Equivalence demands that we establish connections among diverse movements without forcing them to melt into a unitary identity or agenda. The case of Los Angeles homeowners reminds us how difficult such connections are to establish. While there was the potential for homeowners' groups to act in concert with other movements in Los Angeles around an equivalent interest in inhabitance-not-property, the differences among them, rooted primarily in deep-seated ethnic distrust, prevented such links from even being conceived, much less acted on. The possible network of equivalence that could have developed did not. Had it developed, even nascently, it would very likely have had trouble sustaining itself. Differences like Anglo/Latino, affluent/not, and native/immigrant are the dominant contrasting identities in Los Angeles politics. They would have made it a challenge to maintain a shared commitment to inhabitance and democracy. But that challenge is precisely the one that faces movements against neoliberalization and for radical democratization. They must remain different while acting in concert.

While the Duwamish case is quite a lot more promising with respect to equivalence, it nevertheless raises important questions, questions that all movements of equivalence must face. First, just how much difference *can* a coalition hold together? The differences in the case of the DRCC (environmentalist/community activist/Native American advocate) are milder than the differences in the Los Angeles case (low-income Latino bus riders from East Los Angeles versus affluent Anglo homeowners from the Valley). Past what point will such differences make a movement untenable? The second question is just how much difference does a movement *need* to hold together? How wide a coalition is necessary to effectively resist neoliberalization? In general, the larger and more comprehensive the network, the more likely it is to generate political power. And a social mobilization strategy depends on generating power, especially when it pursues a counter-hegemonic project (Harvey 2003). But a larger coalition is also more likely to encompass more irreducible difference in terms of agendas, grievances, and cultural sensibilities. So the two questions are linked: to be successful a movement needs to be as broad and comprehensive as it can, but it does not want to encompass difference that cannot be held together through equivalence. Such questions cannot be answered out of context. They will be strategic questions that each network needs to work through in practice. Equivalence demands a continual and difficult negotiation between sameness and difference.

The need to answer those questions in context points to yet another challenge. A non-essentialist approach like equivalence begins with the assertion that political identities and agendas are not inherent in particular people and groups; they must be *forged* through political organization, mobilization, and struggle. That fact means we must leave room for a variety of political constructions. Political subjects, whether individuals or groups, can construct their identity in a variety of different ways. They can also construct many different understandings of subordination and antagonism, and those understandings can lead to a variety of ways to understand the equivalence that holds a movement together. Therefore, the nature of democracy, or inhabitance, or the right to the city cannot be worked out theoretically, before the fact. It cannot be set in stone. It must constantly be forged and reworked through an iterative process of organization, mobilization, struggle, and critical reflection.

THE PROMISE

In the face of those challenges it is necessary to repeat, over and over, Arundhati Roy's insistence that another world is not only possible, she is on her way. Her claim is a testament of faith, to be sure. It is a faith that spurs on those who face what may seem like an impossibly uphill climb. But in another sense, we can see Roy's claim as simply an empirical fact of history. Human society continually tears down existing relations and rebuilds new ones. Structures endure, but they do not endure forever. Even Fukuyama had to know that "the end of history" was more a rhetorical device than an accurate description of the world. No matter how many times neoliberals repeat that their dominance is the end state, the evolutionary destination of political-economic relations, they can't make it true. All hegemonies present themselves as final resting points. However, they are also, by their very nature, *temporary* arrangements. "Every order," writes Chantal Mouffe (2005, p. 18):

> is the temporary and precarious articulation of contingent practices. The frontier between the social and the political is essentially unstable and requires constant displacements and renegotiations between social agents. Things could always be otherwise and therefore every order is predicated on the exclusion of other possibilities.

The hegemony of neoliberalism, which is about thirty-five years old, is the successor to a Keynesian, corporatist hegemony that itself lasted roughly that long. In the postwar years, when the neoliberal intellectuals of the Mont Pelerin Society were formulating their Statement of Aims, they were a small counter-hegemonic movement that seemed to have little hope of unseating the dominant common sense of Keynesianism. In a 1965 article in *Time* magazine, Milton Friedman himself lamented that "we are all Keynesians now" (Friedman 1965). Richard Nixon repeated the remark even as late as 1971 (Ponnuru 1999). But of course the Keynesian hegemony, as with all hegemonies, was temporary. It collapsed (though, as we have seen, Keynesian structures did not vanish) and neoliberalism rose to prominence. We should expect that neoliberalism will meet the same fate. That is because, in part, hegemonies collapse under their own weight. As Mouffe (through Gramsci) reminds us, hegemonies are active political projects to impose some particular understandings and practices on the world and exclude others. They take enormous political resources to maintain. Any internal contradictions tend to fester, grow, and destabilize the delicate balance. Keynesianism certainly labored under a range of contradictions, and neoliberalism is no different. Just to take one example we have already seen, as neoliberalization exacerbates social inequality it undermines the liberal-democratic requirement that citizens are formal political equals. And so neoliberalization undermines the legitimacy of the very democratic assumptions on which its political legitimacy depends.

Moreover, hegemonies are always not only temporary, but also partial. Neoliberalism did not entirely wipe away Keynesian practices, it suppressed them. Neoliberalization is a process whereby neoliberal ideas are layered on top of existing relations and become dominant relative to previous ideas. That does not mean previous logics and structures disappear. Rather some are eliminated, some are left dormant, and some are woven in to the new policy ensembles of neoliberalization. And that complex process varies from place to place and issue to issue. So even when neoliberal ideas are hegemonic, neoliberalization will always entail a never-completed struggle to tamp down previously dominant logics, values, and practices. We saw in the Duwamish case, for example, how old provisions for public participation, a legacy of Johnson's Great Society initiatives, endure in federal law governing Superfund, and they open up important opportunities for community groups to make an impact on the cleanup. The inherently partial nature of hegemonies, then, means that political opportunities for social movements are always present, even if they vary in their number and quality (Tarrow 1996; Tilly 1984; Meyer and Staggenborg 1996).

Lastly, and perhaps most importantly, hegemonies are always resisted. They are resisted in the sense that still-existing interests from the old order will continually try to reassert their former dominance. Urban planners, for example, even as they operate under the dominant logic of privatization, retain an explicit and deep commitment to *public* solutions to urban problems. Hegemonies are also resisted because groups that are disadvantaged by the dominant order will organize to resist it and to pursue alternatives. It should come as no surprise, for example, that social movements arise to resist the gentrification and displacement that inevitably result when decision-makers prioritize the maximization of exchange value. In other words, resistance is *a necessary part* of the fabric of all hegemonies. And it is the force that does the work. Hegemonies are temporary and partial, and they creak under their internal contradictions, but it takes active, organized, and committed resistance to bring about a collapse.

So, as daunting as the hegemony of neoliberalism might appear from inside its dominion, it is entirely possible to resist it. Counter-hegemonic movements are not only possible but inevitable. Neoliberalism will not be dominant forever; other hegemonies will eventually take its place. But that way of seeing hegemony should breed hope, not complacency. We only know that neoliberalism will not be eternal. We do not know how long it will last or what will replace it. It may continue to immiserate people for another fifty years. Or the alternative, when it comes, may be insufficiently democratic. It may even be less democratic than neoliberalism. Therefore, by no means can counter-hegemonic movements for radical democracy wait around for the collapse. They must actively undermine neoliberalization and advocate for a more democratic alternative.

And in fact they are doing precisely that. Still more reason for optimism is that the world is teeming with movements actively engaged in building counter-hegemonic projects. Part of the intellectual and political task of fostering such democratic movements, therefore, is necessarily empirical. We must discover, learn from, and build on actual movements, such as that on the Duwamish, that are forging networks of equivalence to resist neoliberalization. We must understand movements that are successful, as well as those that fail. It is all very well (and very important) to work out elegant concepts like counter-hegemony and networks of equivalence. However, there is always a danger with such ideas that they appear on paper to hold the solution but are difficult to mobilize in practice. Figuring out how to actually *build* networks of equivalence requires that we study the builders. They exist, and they are having success. It is critical to pay close attention and learn from them.

One lesson the Duwamish case highlights is that neoliberal dominance is riven with cracks. There are significant opportunities for mobilized groups to challenge existing assumptions, structures, and practices, and to propose new ones. Taking advantage of those opportunities requires that such groups be creative, thoughtful, patient, and organized. They must actively seek out the openings. They must creatively discover and invent the tools needed to exploit the cracks, to widen them, and then squeeze into them. As they progressively breach the façade, they both destabilize it and open up opportunities to advance alternatives to neoliberalism. As neoliberal assumptions are undermined, radical democratic ones become ever more possible. It becomes increasingly easy to establish alternative assumptions, construct alternative structures, and pursue alternative practices. It is simultaneously a process of destruction and creation. We might imagine the gains of the Duwamish case being applied productively to other Superfund sites. Many other sites lack the same broad and successful mobilizations for democratization, but claims could be made that use the Duwamish as precedent. Existing mobilizations could consciously mirror the networking in the Duwamish case, working to build equivalence around inhabitance. They could claim the same kinds of access and influence to Superfund decisions that the EPA has granted to the DRCC. And other movements could learn from the DRCC how to carry off the difficult feat of overspilling the channels, of playing the game in order to change it. In that way, by learning from each other and by transferring innovative strategies of resistance, networks of grassroots mobilization can, taken together, transform the structures and assumptions that constrain them. Those transformed structures can, in turn, enable greater organization and mobilization. That is the model being employed in the Brazilian case, as social movements in the *favelas* have prompted legal codification of the right to the city, and that codification opens new opportunities for collective claims. Mobilization and structural transformation are therefore not mutually exclusive political strategies, but two faces of the same approach (Sites 2007; Peck and Tickell 2002a). Paying attention to empirical examples like the Duwamish and the Brazilian *favelas* gives flesh and bone to Laclau and Mouffe's call for movements on the left to challenge the hegemony of neoliberalism and elaborate a credible alternative to it. Laclau and Mouffe exhort us "back to the hegemonic struggle." Mostly what that requires is that we pay attention to the hegemonic struggle that is already taking place.

However, I want to reiterate that we must maintain a *dialogue* between theory and practice. By stressing the lessons to be learned from concrete struggles,

I do not mean at all to say that existing theoretical ideas—about democracy or the right to the city or inhabitance—cannot participate meaningfully in those struggles. They can provide useful linkages among apparently disparate movements, or new ideas about strategy, or a wider analysis for those absorbed in complex everyday politics like the Duwamish cleanup. It can be invaluable, for example, to recognize the Lower Duwamish Waterway Group as a characteristic neoliberal governance structure, to be able to see their agenda as part of a larger neoliberal world view, to use a thoughtfully elaborated concept (like equivalence) to help articulate what the DRCC is trying to achieve, or to offer lessons from other movements that have faced similar challenges. It can be invaluable to mobilize concepts like the right to the city and inhabitance, with their multiple and deep intellectual roots, to give shape and explicit articulation to a movement's existing but latent sense of shared purpose. In building concrete networks, then, there is a very real and important role for more abstract analyses, for theoretical inspiration, and for detailed political and economic history. A non-essentialist approach insists only that such analyses cannot *determine* the politics—theorists can and should voice their perspective in an effort to help *shape* the politics in particular ways.

Accordingly, I have tried to elaborate a set of democratic attitudes that can be useful in conceiving ways to resist neoliberalization and imagine more democratic urban futures. Given that those attitudes point us toward networks of equivalence, I think the right to the city and its claims to inhabitance offer much potential as a set of ideas that networks can take up and use in their particular context. That potential inheres in the fact that the right to the city as I have presented it encompasses many of the same tensions as the politics of equivalence. The right to the city has the potential to be both malleable and directed, to bring together different groups around a coordinated agenda. On the one hand, in order for inhabitance to be a useful basis for equivalence, it needs to offer quite a lot of malleability. It needs to be adaptable to different political contexts, and to different combinations of movements seeking to construct equivalence. Different groups inhabit the city in different ways, and so each will require different ways to imagine inhabitance and what it means to claim a right to the city. In the case of the DRCC, a right to inhabit can mean a right to subsistence fishing, or to play in one's front yard without fear of cancer, or to preserve cultural-ecological sites important to one's tribal history, or to enjoy the peace of the riverbank on a Saturday, or to migrate upstream through healthy ecosystems. Those different ways of inhabiting can come into conflict, as when, for example, at one hotspot LDWG offered two different plans, one with greater riverfront access for residents and the other with more complete ecological restoration for environmentalists.

Because of the ever-present danger of network dissolution, then, the right to the city cannot be infinitely malleable. I argue that it should always retain a sense, which I think is clearly present in Lefebvre, that inhabitance, at its core, is an idea that stands against ownership, commodification, and space-as-property. The different ways people inhabit the city cannot be reduced to each other. However, people can construct a shared sense that they are all equivalently opposed to space as an owned commodity. And the DRCC demonstrates that such a construction is possible. They have made a conscious effort to see different ways to inhabit as cut from the same cloth, and to see that cloth as distinctly different from a cloth woven from property rights and exchange value. They see perfectly well that they must act together to have any kind of impact on the cleanup. At the same time, they understand the error of reducing the agenda of the DRCC to one particular conception of inhabitance. Such reduction would both marginalize some members of the coalition, and, perhaps more importantly, would diminish the coalition's kaleidoscopic breadth, a breadth that is a primary reason EPA–Ecology and LDWG takes them so seriously.

So, as part of the dialogue between theory and practice that I advocate, I have framed my conception of the right to the city by both drawing on Lefebvre, and by imagining what kind of right to the city could best make possible a group like the DRCC. My conception is intended to make room for the many different ways networks might construct equivalent notions of inhabitance. However, it insists on a notion of inhabitance that is, at least in part, always in opposition to the city as owned property. Without that insistence, we can fall into the trap of embracing any movement of inhabitants, even if they do not claim a right to inhabitance. In the case of Los Angeles homeowners, inhabitants can make a claim for a privileged group to preserve its privilege over and above other inhabitants as well as property owners. In that case, we are not really talking about the right to the city as Lefebvre conceives it, and we are not really talking about resistance to neoliberalization in any meaningful form.

So what this book has offered is merely the beginning of the dialogue: a theoretical opening statement that is then developed in the context of concrete politics. The dialogue must be continued as a part of the larger project to create, nurture, and sustain networks of equivalence against the neoliberalization of cities, and for their radical democratization.

BEYOND THE CITY, BEYOND NEOLIBERALIZATION

While it is important to narrow one's focus and develop an in-depth analysis of the particular project of democratizing the neoliberal city, it is also critical to remember what such a focus leaves out. This book considers only democratizing *cities* and resisting *neoliberalization*. That focus should not be mistaken for a claim that the urban is more important than other scales or forms of settlement, or that neoliberalization is more important than other forms of social subordination. The book's project should be taken as part of a larger project to radically democratize social life. So in closing I want to suggest how we might think beyond the urban and beyond neoliberalization.

I think the right to the city can, ironically, help us understand the city in broader context. Lefebvre, along with many others, suggests that there is something magical about the city, that its centrality houses and in fact nurtures the very best in us, that it makes possible the highest achievements humanity is capable of (Lefebvre 1996; Soja 2000; Aristotle 1962). According to that perspective, a right to the *city* is particularly important because it represents a claim for access to the best humans have to offer. But if we set aside for a moment Lefebvre's argument about cities, we can also see the right to the city in a way that allows us to generalize it beyond the city. We can insist that the core of the right to the city is more generally the right to inhabit space, a right opposed to the rights of property and profitability. Conceived of that way, the right to inhabit could encompass the right to the city, but it need not be limited by it. The urban scale need not be more important than either smaller or larger scales. Democratic movements at the neighborhood scale, or national or global scale, can claim the right to inhabit space as a way to resist neoliberalization at a variety of scales. Similarly, and no less importantly, inhabitants in rural areas can claim a right to inhabit space that can have all sorts of impacts, for example on food production and animal husbandry. Land in rural areas around the world is increasingly owned and commodified by industrial-capitalist agricultural production in ways that have profound political-economic and environmental consequences. Claiming a right to inhabit rural space, to use it in ways not primarily determined by the profit imperative, would be one way to unequivocally challenge the neoliberalization of food production, and to pursue its democratization.

And that more general tack would not deny the key role that cities play in capitalist production or the role they have played in neoliberalization. Cities are certainly important strategic sites in which to claim a right to inhabit space. Resisting neoliberalization in places that are particularly important to its project (as laboratories, as beachheads, as critical sites of production) is of course essential to resisting it more generally. However, if we understand the right to *inhabit* as the core of the political claim the right to the city is making, we can see democratizing cities as a necessary, but not sufficient, element of a wider project to resist neoliberalization. We should conceive the project, therefore, as claiming a right to inhabit *in the city*, rather than a right *to the city* per se. While this book has to do with making such claims in the urban context, and while I am seduced by the idea that the city incubates the highest human achievement, I think it is counter-productive to say that the city is *necessarily* more important to resisting neoliberalization than other scales and other forms of settlement. Given the political, economic, and environmental importance of food production, I am not even sure it is right to say that the city is *strategically* more important than rural areas. It is, however, entirely right to say that the city is *a* strategically critical place where we must claim a right to inhabit space. And so of course I am not saying that cities are unimportant, only that they should be seen as strategic sites in a wider movement to claim the right to inhabit space.

Equally critical, and perhaps more difficult, is the need to see neoliberalization as only one hegemony among many that democratization can resist. Here again I think a claim to inhabitance can be extremely useful. Inhabitance has the potential to bring resistance to neoliberalization together with resistance to other forms of oppression. As we know, neoliberalization is not the only form of oppression that confronts urban dwellers. Racism, patriarchy, and heteronormativity, just to name three, are shot through the fabric of urban space. We must resist the privilege and oppression associated with property rights and commodified urban space, but we must also continue to resist the privilege and oppression associated with racial difference and racialized urban space. And gender privilege and gendered urban space. And sexual privilege and sexualized urban space. It is clear that within each of these other forms of oppression, the right to the city can and does offer a way to resist. It is not hard to imagine, for example, how a democratic movement among those disadvantaged by racial segregation could claim the right to fully inhabit the city. The right to inhabit in that case would be constructed as resistance to racial privilege (rather than as resistance to property rights). The same kind of claim could be made by democratic movements against patriarchy or heteronormativity. A right to inhabit urban space certainly includes the right for women to move in urban space without fear of sexual assault, or the right to show same-sex affection without the fear of intimidation or physical attack. As a result, those who have pursued the right to the city in practice have given explicit attention to the many different forms oppression and exclusion can take (e.g. UNESCO 2006). The UNESCO group, for example, is particularly concerned with the issue of religious discrimination and inter-faith tolerance, a much more acute problem in cities in the global South. So the right to the city cannot be *only* about resistance to neoliberalization.

The politics I have explored in this book have been limited to constructing viable networks of equivalence to resist neoliberalization. Those networks construct inhabitance as different-from-ownership, and orchestrate their diverse agendas around that equivalent understanding. However, could we use the right to inhabit as the basis for networks of equivalence *across* forms of oppression? Is inhabitance elastic enough to allow anti-racist movements to act in concert with anti-neoliberal, anti-patriarchy, or anti-heteronormative ones? Could they construct a cohesive movement around the right to inhabit space that makes claims for radical democratization of the city? There is nothing, either in Lefebvre or in the right to the city I have presented, that would prevent such networks. They are possible, in theory. However, just how stable and coherent could they be? They would face the same kind of tensions all such networks face—i.e. the broader they are the more power they have, but the harder they are to keep together. Because such networks would be so broad and encompass such manifest difference, their constant concern would be cohesion. They would likely be tempted toward reduction, toward favoring a particular conception of the right to inhabit (e.g. as different-from-property-rights or as different-from-racial-privilege).

And that kind of reduction need not even be a conscious claim. That is, one need not argue explicitly that a certain way to understand inhabitance is better or more politically important than others. (That of course is the error that some Marxists and socialists have made in the past: the explicit argument that class is a more important basis for political mobilization than other subject positions.) Instead of that explicit claim, one could simply fall into what we might call reduction by habit (Purcell 2002b, 2003b). If we are in the habit of seeing politics in a particular light (e.g. the struggle against neoliberalization), we would also tend to conceive inhabitance in a particular way (e.g. as different-from-ownership). One could certainly read this book that way. It focuses on democratic resistance to neoliberalization, and so it makes it seem like the only way to conceive of inhabitance is as different-from-ownership. I do not argue explicitly that inhabitance-as-not-ownership is more important than other ways to conceive of inhabitance, but of course my discursive emphasis can easily give that impression. The same can happen in practice: a movement can form a network that explicitly incorporates and values different ways to conceive of inhabitance, but nevertheless finds itself most often struggling against a particular form of oppression. Some members of the DRCC, for example, are involved with anti-racist politics, and race and racism are very much factors in the cleanup. However, the politics of the cleanup are such that the day-to-day struggle is against a land-as-property view of the river. DRCC's agenda for the river, as a result of the accretion of their political practice, thus becomes "not-property" more than anything else. They do not consciously devalue questions of race or gender or sexuality, but their everyday practice tends to leave those to the side. As a result, members who conceive of inhabitance primarily as different-from-racial-privilege may begin to question their membership in the network, wondering if it is worth their time and effort. So, even if we have the best of intentions, we can unintentionally reduce difference to one particular conception. When we do, it undermines equivalence because it moves actors toward an *identical* understanding of their agenda, rather than an equivalent one.

However, while there is reason to be sober about the prospects of broader networks, there is also, always, reason for optimism. For that, we must again return to the ground. In everyday urban life, categories like race or class or gender are not separate but intertwined in complex ways. Everyone's everyday life is cross-cut with many different subject positions. They are difficult to pull apart and treat as distinct entities. By extension, actually existing movements generally do not encounter one and only one form of oppression. They are rarely purely anti-neoliberal, or anti-racist, etc. The politics they engage in commonly involve a mixture of race, class, gender, sexuality, etc. Environmental justice movements like the Community Coalition for Environmental Justice, a DRCC member, understand fully that race and class are bound up together in their struggle for healthy environments in the Duwamish cleanup. For another DRCC member, the Duwamish Tribe, issues of historical injustice and disrespect are tied intimately to the pollution and cleanup of riverine ecosystems. For the Duwamish Tribe straightening the river and seeing its banks as property to be sold on the market is entirely inseparable from a history of White conquest and corrosive racial privilege. So everyday experience can be of great help. At the same time, there is no guarantee that such everyday experience will be explicitly constructed equivalently by inhabitants. In order to forge a more inclusive community vision for Terminal 117, DRCC attempted to bridge differences between the Anglo and Latino populations in the South Park neighborhood. They were sincere but not very successful, and Spanish-speaking residents were only minimally included in the process. The gulfs were more than linguistic; they were cultural and class-based as well. While the DRCC has been less successful with that challenge, the point is that such complex and interwoven differences within a coalition are nothing new to most networked movements. Rather they are challenges and opportunities that movements confront all the time. Forging an equivalent notion of inhabitance that ranges across multiple subject positions and multiple forms of oppression, therefore, would not necessarily mean creating those connections out of the ether. Rather it would often require only a conscious effort to bring such connections to the front of people's minds, to articulate together the many different ways that the right to inhabitance is threatened, and how democratic networks can mobilize to defend it.

Again, of course, an anti-essentialist politics means that the particular ways those differences are brought into concert will vary by context. For some networks, the pressure toward reduction will be greater than for others. Some networks will have more experience with intertwined differences than others, or, more likely, most movements will have more experience with some kinds of difference than with other kinds. The politics in each case will be contingent and impossible to model *a priori*. However, each movement will need to confront a similar political task: to draw together a network that organizes different agendas and experiences into equivalence, without reducing them to identicalness. The best way to discover how we might achieve such a feat, I have argued throughout this book, is to pay close attention to those movements that face, and overcome, the challenge every day.

DEMOCRATIZATION AGAINST NEOLIBERALIZATION

Even if we must continually remember that wider context, I hope this book has made the case that there is plenty to occupy us in the specific project to mobilize democratic movements to resist neoliberalization in the city. Part of the effect of neoliberalism's hegemony is to squeeze out all logics other than its own. Claiming a right to the city in that context is a challenge. Moreover, neoliberalization also tends to absorb alternative logics and shape them to its ends. Democratic movements for the right to the city must continually reiterate their fundamental opposition to neoliberalization. They must offer a vision of radical democratization and equalization that stands unequivocally against neoliberalization. They must offer a radical alternative to liberal democracy as it is currently being practiced. They must also, I have argued, offer a radical alternative to mainstream-alternative deliberative and participatory democratic practices. The danger of absorption into neoliberal hegemony must remain a paramount consideration.

But the challenges can't paralyze us. The events in Seattle in 1999 made clear that a movement for a democratic global economy is underway. It helped slow the march of neoliberalization in Cancun, when the Fifth Ministerial Conference of the WTO failed to reach agreement on how to further liberalize global trade. Democratic movements in cities are also standing against neoliberalization and proposing alternative visions of the urban future. I have highlighted the DRCC, but similar movements—for affordable housing, for a living wage, against displacement, for adequate public transportation, for environmental justice, and even specifically, as in Brazil or the UK or Canada or Los Angeles, for "a right to the city"—can be found around the globe. They are getting on with the business of forging more just, more humane, and more democratic cities. They are engaging the hegemonic struggle. That struggle will be difficult, to be sure, but it will eventually result in the end of neoliberal hegemony. That hegemony labors under too many contradictions, and it faces too much organized resistance. It cannot endure forever. What is less clear, however, is what will replace it. Establishing a new hegemony of radically democratic urban political economies is, to say the least, daunting. It will require great effort, hope, and faith. But it is possible. And it has already started. So let's get back to work.

NOTES

CHAPTER 1

1 Giroux 2004.
2 While the costs of coordinating activity across distance have been greatly reduced by advances in transportation and communication technology and the progressive elimination of restrictions on international commerce, nevertheless transcending distance still has significant costs, and so regional clustering has become increasingly common.
3 For what is perhaps the most thorough exposition of the neoliberal policy agenda, see Brenner and Theodore (2002).
4 *Laissez-faire* literally means to "leave to do" and *aidez-faire* means to "help to do."
5 This new preference is typical of a wider neoliberal shift in state spending from welfare to aidez-faire spending. I explore the latter in detail below.
6 This appalling story is increasingly applicable to governance in public universities as well. Department chairs are urged to become more entrepreneurial, because, they are told, public funding isn't forthcoming.
7 Of course "the people" is an extremely complex term and there is much debate about its meaning. I explore that complexity in much detail below.
8 Of course they don't mention the pro-free-market authoritarianism of Pinochet or the Shah of Iran or, it is worth mentioning, Paul Bremer in Iraq (see Klein 2004).

CHAPTER 2

1 Drawing on Mouffe (1999, 2002, 2005), I am making an implicit distinction between agonism and antagonism here, which I develop in detail below.
2 I am rejecting, therefore, an anarchist position that the state apparatus is necessarily antithetical to democracy and that radical democratic movements must by definition oppose the state in all its forms.
3 In what follows, my goal is to characterize the mainstream notion of liberal democracy. I do not include here the ideas of those who wish to reclaim some

tenets of liberal democracy in order to radically democratize society (e.g. Mouffe 1993; Bobbio 2005).

4 It is worth being clear that throughout the book, when I use the term "polity," I mean it in its broadest sense, a political community. For liberal democrats, because politics and the state are so closely tied, polity and political community are often understood more narrowly to mean the formal public sphere governed by the state. If we understand the political more broadly, however—with participatory democrats, revolutionary democrats, and radical pluralists—to mean all relations of power, then a political community encompasses quite a lot more than just the state. It may of course include the state, but state relations may or may not make up a majority of a polity's political relations.

5 Republicanism, of course, goes back to the Greeks (primarily Aristotle) and Romans (Cicero), and to Machiavelli as well, but the strong civic republicanism discussed here is inspired primarily by Rousseau.

6 Thus the tradition of direct democracy, in the sense that it insists on citizens being directly involved in decision-making, has important overlaps with participatory democracy.

7 In Aristotle's time this might have been practicable because so much of the population (women, workers, slaves) were excluded from citizenship; they did the necessary productive labor so that "citizens" were free to participate in politics.

8 Aristotle (1962, 1289a) offers a very similar division, although he called this last form of power "polity," by which he meant the rule of all by the many for the *common* good, and contrasted it with democracy, by which he meant the rule of all by the many in pursuit of their *own* interest. For Aristotle "the many" meant the poor, since he believed they would always outnumber the wealthy.

9 They call that current regime "Empire," and devote an entire book to analyzing it (2000). In sum, Empire is a new form of global political-economic sovereignty that is made up of hybrid and decentered forms of power, but which nonetheless ensures a stable context in which capital can operate.

10 I should point out here that while Fraser's critique is of the liberal notion of the public sphere, her formal target is Habermas' conception of the public sphere, and so her critique can without too much modification be applied to most deliberative democrats as well.

CHAPTER 3

1 This example is hypothetical, but firmly grounded in South Park's real political and environmental history.

2 Some also express concern from a more communitarian or civic-republican perspective. They fear talk of individuals' *rights* squeezes out talk of individuals' *responsibility* to a wider community (e.g. Glendon 2004).

3 To be clear, none of this is to say that non-urban places are not important to capitalism or neoliberalization or democratic resistance. They are. My claim is only that the urban has played an extremely important role in neoliberalization, and so democratic resistance should take the urban seriously.

4 My reading of Lefebvre has been in both French and English, but it relies more on the English than on the French. In Lefebvre (1996) there is a full translation of *Le droit à la ville* and a translation of most of *Espace et politique*. Where the French is untranslated, of course, I have relied on the French.

5 This line of thinking is particularly evident in the recent United Nations initiative (UNESCO 2006).

6 It might also be understood as resistance to other spatial hegemonies as well (e.g. the dominance of straight space (Brown 2000)). That is, opposition to space-as-property is a necessary way to conceive of the right to inhabit, but not the only one. See the book's conclusion for more reflections on politics "beyond neoliberalization."

7 It is worth mentioning that if anyone could have pulled off such a feat it is Lefebvre. He seemed to be almost eerily able to extrapolate how the events of the 1960s would evolve into the neoliberalization we live with today (Lefebvre 2003, p. 6).

8 Another metaphor that is sometimes used here is that of the rhizome, a "nonhierarchical and noncentered network structure," as Hardt and Negri (2000, p. 299) have it. Their conception draws principally on the work of Deleuze and Guattari (1987).

CHAPTER 4

1 This second sense is particularly true of the Port.

2 One challenge the DRCC faces is that EPA and Ecology, when they operate separately, often apply different requirements to a site's cleanup plan. Standards, strategies, and community involvement can therefore vary from site to site. The DRCC must continually adapt to shifting conditions, which, while it can open up opportunities, burdens them with the need to reassess the situation at each site.

3 Council members are all elected at large, so each member has a citywide constituency. So for each council member South Park is only a small part of their constituency.

References

Adler, J. (1998) A New Environmental Federalism: Environmental Policymakers Are Increasingly Turning to the States for Solutions to Today's Environmental Problems. *Forum for Applied Research and Public Policy* 13(4): 55–61.

Aglietta, M. (1979) *A Theory of Capitalist Regulation.* London, New Left Books.

Agnew, J. (1994) The Territorial Trap: The Geographical Assumptions of International Relations Theory. *Review of International Political Economy* 1(1): 53–80.

Altshuler, A. and D. Luberoff (2003) *Mega-Projects: The Changing Politics of Urban Public Investment.* Washington, DC, Brookings Institution Press.

Amin, A. (ed.). (1994) *Globalization, Institutions and Regional Development.* Oxford, Oxford University Press.

Appiah, K. (2004) *The Ethics of Identity.* Princeton, Princeton University Press.

Aristotle (1962) *The Politics of Aristotle.* New York, Oxford University Press.

Aristotle (2004) *The Nicomachean Ethics.* New York, Penguin Classics.

Aronowitz, S. (2001) *The Knowledge Factory: Dismantling the Corporate University and Creating True Higher Learning.* Boston, Beacon Press.

Arrow, K. (1951) *Social Choice and Individual Values.* New York, Wiley.

Associated Press (2006) Rice Sees Triumph of Democracy in Chile. *The Jamaica Observer.* March 12.

Autant-Bernard, C., V. Mangematin and N. Massard (2006) Creation and Growth of High-Tech SMEs: The Role of the Local Environment. *Small Business Economics* 26(2): 173–87.

Bachrach, P. (1967) *The Theory of Democratic Elitism: A Critique.* Boston, Little, Brown.

Baiocchi, G. (2003) Participation Activism and Politics: The Porto Alegre Experiment. *Deepening Democracy.* A. Fung and E. Wright (eds). New York, Verso: 45–76.

Balibar, E. (1999) Is European Citizenship Possible? *Cities and Citizenship.* J. Holston (ed.). Durham, NC, Duke University Press: 195–215.

Barber, B. (2004) *Strong Democracy.* Berkeley, University of California Press.

BBC (1998) Basque Elections "Victory for Democracy". http://www.bbc.co.uk. October 26.

Bell, D. (1993) *Communitarianism and Its Critics*. Oxford, Oxford University Press.

Benhabib, S. (1994) Deliberative Rationality and Models of Democratic Legitimacy. *Constellations* 1(1): 25–53.

Benhabib, S. (1996) Toward a Deliberative Model of Democratic Legitimacy. *Democracy and Difference: Contesting Boundaries of the Political*. S. Benhabib (ed.). Princeton, Princeton University Press: 67–94.

Berlin, I. (1969) *Four Concepts of Liberty*. Oxford, Oxford University Press.

Bernholz, P. (2000) Democracy and Capitalism: Are They Compatible in the Long-Run? *Journal of Evolutionary Economics* 10(1–2): 3–16.

Blakely, E. and M. Snyder (1997) *Fortress America: Gated Communities in the United States*. Washington, DC, Brookings Institution Press.

Bobbio, N. (2005) *Liberalism and Democracy*. London, Verso.

Bohman, J. and W. Rehg (eds) (1997) *Deliberative Democracy*. Cambridge, MA, MIT Press.

Bowles, S. and H. Gintis (1986) *Democracy and Capitalism: Property, Community, and the Contradictions of Modern Social Thought*. New York, Basic Books.

Boyarsky, B. (1997a) If Arena Deal Is So Good, Show Us the Leases. *Los Angeles Times*. June 23, p. B1.

Boyarsky, B. (1997b) Wall of Secrecy Surrounds Key Part of Arena Deal. *Los Angeles Times*. June 30, p. B1.

Brenner, N. (1997) State Territorial Restructuring and the Production of Spatial Scale: Urban and Regional Planning in the FRG, 1960–1989. *Political Geography* 16: 273–306.

Brenner, N. (1999) Globalisation as Reterritorialization: The Re-Scaling of Urban Governance in the European Union. *Urban Studies* 36(3): 431–51.

Brenner, N. (2000) The Urban Question as a Scale Question: Reflections on Henri Lefebvre, Urban Theory and the Politics of Scale. *International Journal of Urban and Regional Research* 24(2): 361–78.

Brenner, N. (2001) The Limits to Scale? Methodological Reflections on Scalar Structuration. *Progress in Human Geography* 25(4): 591–614.

Brenner, N. (2005) *New State Spaces: Urban Governance and the Rescaling of Statehood*. New York, Oxford University Press.

Brenner, N. and N. Theodore (2002) Cities and the Geographies of "Actually Existing Neoliberalism". *Antipode* 34(3): 349–79.

Brenner, R. and M. Glick (1991) The Regulation Approach: Theory and History. *New Left Review* 188: 45–120.

Brown, J. C. and M. Purcell (2005) There's Nothing Inherent About Scale: Political Ecology, the Local Trap, and the Politics of Development in the Brazilian Amazon. *Geoforum* 36: 607–24.

Brown, M. (2000) *Closet Space*. New York, Routledge.

Brownhill, S., K. Razzaque, T. Stirling and H. Thomas (1996) Local Governance and the Racialization of Urban Policy in the UK: The Case of Urban Development Corporations. *Urban Studies* 33: 1337–55.

Bryan, F. (2003) *Real Democracy: The New England Town Meeting and How It Works*. Chicago, University of Chicago Press.

Burgess, J. and C. Harrison (1998) Environmental Communication and the Cultural Politics of Environmental Citizenship. *Environment and Planning A* 30: 1445–60.

Buroni, T. (1998) A Case for the Right to Habitat. Seminar on urban poverty, Rio de Janeiro, May.

Bush, G. (2005a) President Discusses War on Terror at National Endowment for Democracy. Washington, DC, The White House.

Bush, G. (2005b) State of the Union Address. Washington, DC, The White House.

Butcher, T. (2006) Surprise Triumph for Democracy over Guns. *The Daily Telegraph*. January 26, p. 17.

Calhoun, C. (1988) The Radicalism of Tradition and the Question of Class Struggle. *Rationality and Revolution*. M. Taylor (ed.). New York, Cambridge University Press: 129–75.

Castells, M. (1977) *The Urban Question: A Marxist Approach*. Cambridge, MA, MIT Press.

Castells, M. (1983) *The City and the Grassroots*. London, Edward Arnold.

Ceis, T. (2004) Deputy Mayor's Comments to City Council. Seattle, Office of the Mayor, July 13.

Ceraso, K. (1999) Seattle Neighborhood Planning. *Shelterforce Online* 108.

Chan, S. (2006) Meeting Basic Needs of Homeless Is Goal of New Downtown Facility. *Seattle Times*. May 17, p. B2.

Chandler, D. (2002) *From Kosovo to Kabul: Human Rights and International Intervention*. London, Pluto Press.

Christopherson, S. and M. Storper (1989) The Effects of Flexible Specialization on Industrial Politics and the Labor Market: The Motion Picture Industry. *Industrial and Labor Relations Review* 42(3): 331–47.

Cities for Human Rights (1998) Conference Held in Barcelona, October 17.

City & Shelter *et al.* (no date) The European Charter for Women in the City. Brussels, Commission of the European Union (Equal Opportunities Unit).

City Council (2003) Resolution 30610. Seattle, City of Seattle, June 9.

Cohen, J. (1997) Deliberation and Democratic Legitimacy. *Deliberative Democracy: Essays on Reason and Politics*. J. Bohman and W. Rehg (eds). Cambridge, MA, MIT Press: 67–91.

Cohen, J. and J. Rogers (1995) *Associations and Democracy*. London, Verso.

Connolly, W. (1991) *Identity/Difference: Democratic Negotiations of Political Paradox*. Ithaca, Cornell University Press.

Courchene, T. (1995) Glocalization: The Regional/International Interface. *Canadian Journal of Regional Science* 18: 1–20.

Cunningham, F. (2001) *Theories of Democracy*. New York, Routledge.

Dahl, R. (1967) *Pluralist Democracy in the United States*. Chicago, Rand McNally.

Dahl, R. (1985) *A Preface to Economic Democracy*. Berkeley, University of California Press.

Daniel, C. (2001) Participatory Urban Governance: The Experience of Santo Andre. *United Nations Chronicle Online* 38(1).

Davis, M. (1990) *City of Quartz: Excavating the Future in Los Angeles*. New York, Vintage Books.

Davis, M. (2001) *Magical Urbanism: Latino Reinvent the US Big City*. New York, Verso.

Davis, M. (2006) *Planet of Slums*. New York, Verso.

Day, R. (2005) *Gramsci is Dead: Anarchist Currents in the Newest Social Movements*. Ann Arbor, Pluto Press.

de Tocqueville, A. (2004) *Democracy in America*. New York, Library of America.

Deleuze, G. and F. Guattari (1987) *A Thousand Plateaus*. Minneapolis, University of Minnesota Press.

Department of Planning and Development (2005) South Lake Union Neighborhood Plan Update Underway. Seattle, City of Seattle.

Department of Planning and Development (2006a) *Seattle's Central Waterfront Concept Plan*. Seattle, City of Seattle.

Department of Planning and Development (2006b) Visioning Charrette. Seattle, City of Seattle: Available at: http://www.seattle.gov/dpd/Planning/Central_Waterfront/CharretteExhibit/default.asp#team.

Department of Urban Planning and Development (2005) City Council Adopts Comprehensive Plan Amendments. Seattle, City of Seattle.

Derrida, J. (1988) *Limited, Inc*. Evanston, Northwestern University Press.

Dewey, J. (2004) *Democracy and Education*. Mineola, Dover Publications.

Dicken, P. (1998) *Global Shift*. New York, Guilford.

Diers, J. (2004) *Neighbor Power: Building Community the Seattle Way*. Seattle, University of Washington Press.

Dietrich, H. (2003) $250m Eyed to Grow Biotech Here. *Puget Sound Business Journal*. November 28.

Dikec, M. (2001) Justice and the Spatial Imagination. *Environment and Planning A* 33: 1785–805.

Douglass, F. (1985[1857]) *The Frederick Douglass Papers*. New Haven, Yale University Press.

Downs, A. (1957) *An Economic Theory of Democracy*. New York, Harper.

Downs, D. (1989) *The New Politics of Pornography*. Chicago, University of Chicago Press.

Dryzek, J. (1990) *Discursive Democracy*. Cambridge, Cambridge University Press.

Dryzek, J. (1996) *Democracy in Capitalist Times*. Oxford, Oxford University Press.

Dryzek, J. (2000) *Deliberative Democracy and Beyond*. Oxford, Oxford University Press.

Dumenil, G. and D. Levy (2004) *Capital Resurgent: Roots of the Neoliberal Revolution*. Cambridge, Harvard University Press.

Dunn, S. (2002) Introduction. *The Social Contract and the First and Second Discourses*. S. Dunn (ed.). New Haven, Yale University Press: 1–35.

Dworkin, R. (1983) In Defense of Equality. *Social Philosophy and Policy* 1(1): 24–40.

Dworkin, R. (1984) Rights as Trumps. *Theories of Rights*. J. Waldron (ed.). New York, Oxford University Press: 153–67.

Easterbrook, G. (2005) Future Tense. *The New Republic Online*. January 31.

Eckersley, R. (2004) *The Green State: Rethinking Democracy and Sovereignty*. Boston, MIT Press.

Eisinger, P. (2000) The Politics of Bread and Circuses. *Urban Affairs Review* 35(3): 316–33.

England, M. (2006) *Citizens on Patrol: Community Policing and the Territorialization of Public Space in Seattle, Washington*. Ph.D. Dissertation, Department of Geography, University of Kentucky.

Etzioni, A. (1994) *The Spirit of Community*. New York, Touchstone.

Fainstein, S. (2000) New Directions in Planning Teory. *Urban Rural Review* 35(4): 451–78.

Fainstein, S. (2001) Competitiveness, Cohesion and Governance: Their Implications for Social Justice. *International Journal of Urban and Regional Research* 25(4): 884–8.

Fernandes, E. (2006) Updating the Declaration of the Rights of Citizens in Latin America: Constructing the "Right to the City" in Brazil. *International Public Debates: Urban Policies and the Right to the City.* UNESCO (ed.). Paris, UNESCO: 40–53.

Fishkin, J. (1997) *The Voice of the People.* New Haven, Yale University Press.

Fishman, R. (1987) *Bourgeois Utopias: The Rise and Fall of Suburbia.* New York, Basic Books.

Flyvbjerg, B. (1998a) Empowering Civil Society: Habermas, Foucault and the Question of Conflict. *Cities for Citizens: Planning and the Rise of Civil Society in a Global Age.* M. Douglass and J. Friedmann (eds). New York, Wiley: 185–211.

Flyvbjerg, B. (1998b) *Rationality and Power: Democracy in Practice.* Chicago, University of Chicago Press.

Flyvbjerg, B., N. Bruzelius and W. Rothengatter (2003) *Megaprojects and Risk.* Cambridge, Cambridge University Press.

Fogelson, R. (1967) *Fragmented Metropolis: Los Angeles, 1850–1930.* Berkeley, University of California Press.

Forero, J. (2002) Uprising in Venezuela. *New York Times.* April 13, p. A2.

Forester, J. (1998) Rationality, Dialogue and Learning: What Community and Environmental Mediators Can Teach Us About the Practice of Civil Society. *Cities for Citizens: Planning and the Rise of Civil Society in a Global Age.* M. Douglass and J. Friedmann (eds). New York, Wiley.

Forester, J. (1999) Dealing with Deep Value Differences. *The Consensus Building Handbook.* L. Susskind, S. McKearnan and J. Thomas-Larmer (eds). Thousand Oaks, Sage: 463–93.

Fortune (2006) The Index. *Fortune* 153(7).

Foucault, M. (1988) The Ethic of Care for the Self as a Practice of Freedom. *The Final Foucault.* J. Bernauer and D. Rasmussen (eds). Cambridge, MA, MIT Press: 1–20.

Fraser, N. (1990) Rethinking the Public Sphere: A Contribution to the Critique of Actually Existing Democracy. *Social Text* 25/26: 56–80.

Fraser, N. (1995) From Redistribution to Recognition? Dilemmas of Justice in a "Post-Socialist" Age. *New Left Review* 212: 68–93.

Fraser, N. (1997) *Justus Interruptus: Critical Reflections on the Post-Socialist Condition.* New York, Routledge.

Fraser, N. (2001) Recognition without Ethics? *Theory, Culture & Society* 18(2–3): 21–42.

Fred Hutchinson Cancer Research Center (2004) *The Changing Face of South Lake Union.* Seattle, Fred Hutchinson Cancer Research Center, June 3.

Friedman, M. (1962) *Capitalism and Freedom.* Chicago, University of Chicago Press.

Friedman, M. (1965) We Are All Keynesians Now. *Time.* December 31.

Friedman, T. (2005) *The World Is Flat.* New York, Farrar, Straus and Giroux.

Friedmann, J. (1995) The Right to the City. *Society and Nature* 1(1): 71–84.

Fukuyama, F. (1989) The End of History? *National Interest* 16: 3–18.

Fukuyama, F. (1992) *The End of History and the Last Man.* New York, Free Press.

Fuller, B. (2006) Surprising Cooperation in a World of Difference and Conflict: Water Management in California and Florida. Seattle, WA, Presentation at the Department of Urban Design & Planning, University of Washington, March 9.

Fung, A. (2004) *Empowered Participation: Reinventing Urban Democracy*. Princeton, Princeton University Press.

Fung, A. and E. Wright (2003) Thinking About Empowered Participatory Governance. *Deepening Democracy*. A. Fung and E. Wright (eds). New York, Verso: 3–42.

Gamson, W. (1990) *The Strategy of Social Protest*. Belmont, Wadsworth.

Garber, A. (2007) Viaduct Surface Option Eyed. *Seattle Times*. February 8.

Gastil, J. (1993) *Democracy in Small Groups: Participation, Decision Making, and Communication*. Philadelphia, New Society.

Geiger, R. (2004) *Knowledge and Money: Research Universities and the Paradox of the Marketplace*. Palo Alto, Stanford University Press.

Genet, C. (1997) Quelles Conditions Pour La Formation Des Biotechnopoles: Une Analyse Dynamique. *Revue d'Economie Regionale et Urbaine* 3: 405–24.

Gill, S. (1996) Globalization, Democratization, and the Politics of Indifference. *Globalization: Critical Reflections*. J. Mittelman (ed.). Boulder, Lynne Rienner: 205–28.

Gilmore, R. (2006) *Golden Gulag: Prisons, Surplus, Crisis, and Opposition in Globalizing California*. Berkeley, University of California Press.

Gintis, H. (1972) Consumer Behavior and the Concept of Sovereignty: Explanations of Social Decay. *The American Economic Review* 62: 267–78.

Giroux, H. (2004) *The Terror of Neoliberalism: Authoritarianism and the Eclipse of Democracy*. New York, Paradigm.

Glendon, M. (2004) *Rights Talk*. Northampton, MA, Free Press.

Goodwin, M. (1991) Replacing a Surplus Population: The Policies of the London Docklands Development Corporation. *Housing and Labor Markets: Building the Connections*. J. Allen and C. Hamnett (eds). London, Allen and Unwin.

Goodwin, M. and J. Painter (1996) Local Governance, the Crises of Fordism and the Changing Geographies of Regulation. *Transactions of the Institute of British Geographers* 21: 635–48.

Goonewardena, K. (2003) The Future of Planning at the "End of History". *Planning Theory* 2(3): 183–224.

Gordon, L. and G. Whitty (1997) Giving the "Hidden Hand" a Helping Hand? The Rhetoric and Reality of Neoliberal Education Reform in England and New Zealand. *Comparative Education* 33(3): 453–67.

Gottdiener, M. (1994) *The Social Production of Urban Space*. Austin, University of Texas Press.

Gould, C. (1988) *Rethinking Democracy*. Cambridge, Cambridge University Press.

Graeber, D. (2004) The New Anarchists. *A Movement of Movements: Is Another World Really Possible?* T. Mertes (ed.). New York, Verso: 202–15.

Gramsci, A. (1971) *Selections from the Prison Notebooks*. New York, International Publishers.

Grant Building Tenants Association (2001) Grant Building Saved (for the Moment)! San Francisco, Grant Building Tenants Association.

Green, G. and A. Haines (2001) *Asset Building and Community Development*. Thousand Oaks, Sage.

Grengs, J. (2002) Community-Based Planning as a Source of Political Change. *Journal of the American Planning Association* 68(2): 165–78.

Gutmann, A. and D. Thompson (2004) *Why Deliberative Democracy?* Princeton, Princeton University Press.

Habermas, J. (1984) *The Theory of Communicative Action*. Boston, Beacon Press.

Habermas, J. (1985) *The Theory of Communicative Action, Vol. 2: Lifeworld and System— a Critique of Functionalist Reason*. Boston, Beacon Press.

Habermas, J. (1990) *Moral Consciousness and Communicative Action*. Cambridge, MA, MIT Press.

Habermas, J. (1994) Burdens of the Double Past. *Dissent* 41(4): 513–17.

Habermas, J. (1999) A Short Reply. *Ratio Juris* 12(4): 445–53.

Habermas, J. (2001) Deliberative Politics. *Democracy*. D. Estlund (ed.). Oxford, Blackwell: 107–26.

Harding, A. (1995) Elite Theory and Growth Machines. *Theories of Urban Politics*. D. Judge, G. Stoker and H. Wolman (eds). Thousand Oaks, Sage: 35–53.

Hardt, M. and A. Negri (2000) *Empire*. Cambridge, MA, Harvard University Press.

Hardt, M. and A. Negri (2004) *Multitude: War and Democracy in the Age of Empire*. New York, Penguin.

Hartz, L. (1955) *The Liberal Tradition in America*. New York, Harcourt, Brace & World.

Harvey, D. (1982) *The Limits to Capital*. Oxford, Blackwell.

Harvey, D. (1985) *Consciousness and the Urban Experience*. Oxford, Basil Blackwell.

Harvey, D. (1989) From Managerialism to Entrepreneurialism: The Transformation of Urban Governance in Late Capitalism. *Geografiska Annaler* 71B: 3–17.

Harvey, D. (1996) *Justice, Nature and the Geography of Difference*. Cambridge, MA, Blackwell.

Harvey, D. (2003) A Right to the City. *International Journal of Urban and Regional Research* 27(4): 939–41.

Harvey, D. (2005) *A Brief History of Neoliberalism*. New York, Oxford University Press.

Healey, A. (2006) Vulcan Plans a "Thriving" Mixed-Use Community. *Seattle Daily Journal of Commerce*. October 5.

Healey, P. (1992) Planning through Debate. *Town Planning Review* 63(2): 143–62.

Healey, P. (1997) *Collaborative Planning: Shaping Places in Fragmented Societies*. London, Palgrave.

Hillier, J. (2002) Direct Action and Agonism in Democratic Planning Practice. *Planning Futures: New Directions for Planning Theory*. P. Allmendinger and M. Tewdwr-Jones (eds). New York, Routledge: 110–35.

Hillier, J. (2003) "Agon"izing over Consensus: Why Habermasian Ideals Cannot Be "Real". *Planning Theory* 2(1): 37–59.

Hirst, P. (1994) *Associative Democracy*. Amherst, University of Massachusetts Press.

Hirst, P. and G. Thompson (1995) *Globalization in Question*. Cambridge, Polity Press.

Hirst, P. and J. Zeitlin (eds) (1989) *Reversing Industrial Decline? Industrial Structure and Policy in Britain and Her Competitors*. Oxford, Berg.

Hoffmann, S. (2006) The Foreign Policy the U.S. Needs. *New York Review of Books* 53(13): 60–4.

Hoggett, P. (1987) A Farewell to Mass Production? Decentralization as an Emergent Private and Public Sector Paradigm. *Decentralization and Democracy: Localizing Public Services*. P. Hoggett and R. Hambleton (eds). Bristol, School for Advanced Urban Studies.

Holston, J. (1998) Spaces of Insurgent Citizenship. *Making the Invisible Visible: A Multicultural Planning History*. L. Sandercock (ed.). Berkeley, University of California Press: 35–52.

Honneth, A. (1995) *The Struggle for Recognition: The Moral Grammar of Social Conflicts.* Cambridge, Polity Press.

Howard, D. (1990) Rediscovering the Left. *Praxis International* 10(3/4): 193–204.

Hunter, F. (1953) *Community Power Structure: A Study of Decision Makers.* Chapel Hill, University of North Carolina Press.

Huxley, M. (2000) The Limits to Communicative Planning. *Journal of Planning Education and Research* 19: 369–77.

Huxley, M. (2002) Governmentality, Gender, Planning: A Foucauldian Perspective. *Planning Futures: New Directions for Planning Theory.* P. Allmendinger and M. Tewdwr-Jones (eds). New York, Routledge: 136–54.

Innes, J. (1995) Planning Theory's Emerging Paradigm. *Journal of Planning Education and Research* 14(3): 183–9.

Innes, J. (1996) Planning through Consensus Building: A New View of the Comprehensive Planning Ideal. *Journal of the American Planning Association* 62(4): 460–72.

Innes, J. (1998) Information in Communicative Planning. *Journal of the American Planning Association* 64(1): 52–63.

Innes, J. (2004) Consensus Building: Clarifications for the Critics. *Planning Theory* 3(1): 5–20.

Innes, J. and D. Booher (1999) Consensus Building as Role Playing and Bricolage: Toward a Theory of Collaborative Planning. *Journal of the American Planning Association* 65(1): 9–27.

Innes, J. and D. Booher (2000) Public Participation in Planning. Annual Conference of the Association of Collegiate Schools of Planning, November 2–5.

International Network for Urban Research and Action (1998) *Possible Urban Worlds: Urban Strategies at the End of the 20th Century.* Boston, Birkhauser.

International Network for Urban Research and Action (2003) An Alternative Urban World Is Possible: A Declaration for Urban Research and Action. *International Journal of Urban and Regional Research* 27(4): 952–5.

Isin, E. (2000) Introduction: Democracy, Citizenship and the City. *Democracy, Citizenship and the Global City.* E. Isin (ed.). New York, Routledge: 1–21.

Jaggar, A. (1988) *Feminist Politics and Human Nature.* Totowa, Rowman and Littlefield.

Jawara, F. and A. Kwa (2003) *Behind the Scenes at the WTO: The Real World of International Trade Negotiations.* New York, Zed Press.

Jenkins, C. (1983) Resource Mobilization Theory and the Study of Social Movements. *Annual Review of Sociology* 9: 527–53.

Jessop, B. (1990) *State Theory: Putting Capitalist States in Their Place.* University Park, Pennsylvania State University Press.

Jessop, B. (1993) Towards a Schumpterian Workfare State? Preliminary Remarks on Post-Fordist Political Economy. *Studies in Political Economy* 40(Spring): 7–39.

Jessop, B. (1994a) Post-Fordism and the State. *Post-Fordism: A Reader.* A. Amin (ed.). Malden, Blackwell: 251–79.

Jessop, B. (1994b) The Transition to Post-Fordism and the Schumpterian Workfare State. *Towards a Post-Fordist Welfare State.* R. Burrows and B. Loader (eds). London, Routledge: 13–38.

Jessop, B. (1997a) A Neo-Gramscian Approach to the Regulation of Urban Regimes: Accumulation Strategies, Hegemonic Projects, and Governance. *Reconstructing Urban Regime Theory: Regulating Urban Politics in a Global Economy.* M. Lauria (ed.). Thousand Oaks, Sage: 51–73.

Jessop, B. (1997b) Twenty Years of the (Parisian) Regulation Approach: The Paradox of Success and Failure at Home and Abroad. *New Political Economy* 2(3): 503–26.

Jessop, B. (2000) The Crisis of the National Spatio-Temporal Fix and the Tendential Ecological Dominance of Globalizing Capitalism. *International Journal of Urban and Regional Research* 24(2): 323–60.

Jessop, B. (2002) Liberalism, Neoliberalism, and Urban Governance: A State-Theoretical Perspective. *Antipode* 34(3): 452–72.

Jones, B. and M. Keating (eds) (1995) *The European Union and the Regions*. Oxford, Clarendon Press.

Judge, D. (1995) Pluralism. *Theories of Urban Politics*. D. Judge, G. Stoker and H. Wolman (eds). Thousand Oaks, CA, Sage: 13–34.

Kant, I. (2003) *Perpetual Peace*. Indianapolis, Hackett.

Kearns, A. (1995) Active Citizenship and Local Governance: Political and Geographical Dimensions. *Political Geography* 14(2): 155–76.

Keating, M. (1991) Local Development Politics in Britain and France. *Journal of Urban Affairs* 13: 443–59.

Keating, W. (1997) Cleveland: The "Comeback City": The Politics of Redevelopment and Sports Stadiums Amidst Urban Decline. *Reconstructing Urban Regime Theory: Regulating Urban Politics in a Global Economy*. M. Lauria (ed.). Thousand Oaks, Sage: 189–205.

Keil, R. (2002) "Common-Sense" Neoliberalism: Progressive Conservative Urbanism in Toronto, Canada. *Antipode* 34(3): 578–601.

Kelling, G. and J. Wilson (1982) Broken Windows. *Atlantic Monthly* 249(3): 29–38.

Khor, M. (1999) The Revolt of the Developing Nations. *South–North Development Monitor (SUNS)* 4569(December 8).

Klein, N. (2004) Baghdad Year Zero: Pillaging Iraq in Pursuit of Neocon Utopia. *Harper's Magazine* September: 43–53.

Kofman, E. and E. Lebas (1996) Lost in Transposition: Time, Space, and the City. *Writings on Cities*. E. Kofman and E. Lebas (eds). Cambridge, MA, Blackwell: 3–60.

Kriesi, H. (1995) The Political Opportunity Structure of New Social Movements. *The Politics of Social Protest: Comparative Perspectives on States and Social Movements*. C. Jenkins and B. Klandermans (eds). Minneapolis, University of Minnesota Press: 167–98.

Krumholz, N. (1999) Equitable Approaches to Local Economic Development. *Policy Studies Journal* 27(1): 83–95.

Kymlicka, W. (1990) *Contemporary Political Philosophy: An Introduction*. Oxford, Clarendon Press.

Laclau, E. (1996) *Emancipation(s)*. New York, Verso.

Laclau, E. and C. Mouffe (1985) *Hegemony and Socialist Strategy: Towards a Radical Democratic Politics*. London, Verso.

Laclau, E. and C. Mouffe (2000) Preface to the Second Edition. *Hegemony and Socialist Strategy*. E. Laclau and C. Mouffe (eds). New York, Verso: vii–xix.

Landesman, C. and R. Meeks (eds) (2002) *Philosophical Skepticism: From Plato to Rorty*, Malden, Blackwell.

Lange, L. (2006) Cost to Replace Viaduct Could Rise Drastically. *Seattle Post-Intelligencer*. September 20.

Larner, W. (1997) The Legacy of the Social: Market Governance and the Consumer. *Economy and Society* 26(3): 373–99.

Larner, W. (2000) Neo-Liberalism: Policy, Ideology, Governmentality. *Studies in Political Economy* 63: 5–25.

Larner, W. (2005) Co-Constituting "after Neoliberalism": New Forms of Governance in Aotearoa New Zealand. Towards a Political Economy of Scale, York University, Toronto, February.

Lebowitz, M. (2006) The Politics of *Beyond Capital*. *Historical Materialism* 14(4): 167–83.

Lefebvre, H. (1968) *Le Droit à la Ville*. Paris, Anthropos.

Lefebvre, H. (1973) *Espace et Politique*. Paris, Anthropos.

Lefebvre, H. (1978) *De l'Etat: Les Contradictions de l'Etat Moderne*. Paris, Union Générale de l'Editions.

Lefebvre, H. (1979) Space: Social Product and Use Value. *Critical Sociology: European Perspectives*. J. Frieberg (ed.). New York, Irvington.

Lefebvre, H. (1986) Pour Une Nouvelle Culture Politique. *M: Mensuel, Marxisme, Mouvement* 1.

Lefebvre, H. (1991a) *Critique of Everyday Life*. London, Verso.

Lefebvre, H. (1991b) Les Illusions De La Modernité. *Manières de voir* 13: 14–17.

Lefebvre, H. (1991c) *The Production of Space*. Oxford, Blackwell.

Lefebvre, H. (1996) *Writings on Cities*. Cambridge, MA, Blackwell.

Lefebvre, H. (2003) *The Urban Revolution*. Minneapolis, University of Minnesota Press.

Lefort, C. (1988) *Democracy and Political Theory*. Minneapolis, University of Minnesota Press.

Leitner, H. (1997) Reconfiguring the Spatiality of Power: The Construction of a Supernational Migration Framework for the European Union. *Political Geography* 16(2): 123–44.

Levine, A. (1981) *Liberal Democracy: A Critique of Its Theory*. New York, Columbia University Press.

Lipietz, A. (1992) The Regulation Approach and Capitalist Crisis: An Alternative Compromise for the 1990s. *Cities and Regions in the New Europe*. M. Dunford and G. Kafkalas (eds). London, Belhaven Press: 309–34.

Locke, J. (1988) *Two Treatises of Government*. New York, Cambridge University Press.

Low, M. (2004) Cities as Spaces of Democracy: Complexity, Scale and Governance. *Spaces of Democracy*. C. Barnett and M. Low (eds). London, Sage: 128–46.

Lowy, M. (2003) *The Theory of Revolution in the Young Marx*. Leiden, Brill.

Lowy, M. and F. Betto (no date) Values of a New Civilization. Available at http://www.forumsocialmundial.org.br/download/tconferencias_freibeto_eng.rtf, World Social Forum.

Lummis, C. Douglas (1997) *Radical Democracy*. Ithaca, Cornell University Press.

Luxemburg, R. (1970) *Rosa Luxemburg Speaks*. New York, Pathfinder Press.

McAdam, D. (1996) Conceptual Origins, Current Problems, Future Directions. *Comparative Perspectives on Social Movements*. D. McAdam, J. McCarthy and M. Zald (eds). New York, Cambridge University Press: 23–40.

McCarthy, J. and M. Zald (1977) Resource Mobilization and Social Movements: A Partial Theory. *American Journal of Sociology* 82: 1212–41.

McCarthy, T. (1994) Kantian Constructivism and Reconstructivism: Rawls and Habermas in Dialogue. *Ethics* 105(1): 44–64.

McGuirk, P. (2001) Situating Communicative Planning Theory: Context, Power, and Knowledge. *Environment and Planning A* 33: 195–217.

McKearnan, S. and D. Fairman (1999) Producing Consensus. *The Consensus Building Handbook*. L. Susskind, S. McKearnan and J. Thomas-Larner (eds). Thousand Oaks, Sage: 325–74.

McKenzie, E. (1994) *Privatopia: Homeowner Associations and the Rise of Residential Private Government*. New Haven, Yale University Press.

McKenzie, E. (1998) Homeowner Associations and California Politics: An Exploratory Analysis. *Urban Affairs Review* 34(1): 52–75.

MacLeod, G. (2002) From Urban Entrepreneurialism to a Revanchist City? On the Spatial Injustices of Glasgow's Renaissance. *Antipode* 34(3): 602–24.

MacLeod, G. and M. Goodwin (1999a) Reconstructing an Urban and Regional Political Economy: On the State, Politics, Scale, and Explanation. *Political Geography* 18(6): 697–730.

MacLeod, G. and M. Goodwin (1999b) Space, Scale and State Strategy: Rethinking Urban and Regional Governance. *Progress in Human Geography* 23(4): 503–27.

Macpherson, C. (1973) *Democratic Theory: Essays in Retrieval*. London, Oxford University Press.

Madison, J. (1987[1788]) *The Federalist Papers*. New York, Penguin.

Mansbridge, J. (1983) *Beyond Adversary Democracy*. Chicago, University of Chicago Press.

Mansbridge, J. (1990) Feminism and Democracy. *The American Prospect* 1(1).

Mansbridge, J. (1992) A Deliberative Theory of Interest Representation. *The Politics of Interests: Interest Groups Transformed*. M. Petracca (ed.). Boulder, Westview Press.

Marston, S. (2000) The Social Construction of Scale. *Progress in Human Geography* 24(2): 219–42.

Martin, R. and P. Sunley (1997) The Post-Keynesian State and the Space Economy. *Geographies of Economies*. R. Lee and J. Wills (eds). London, Arnold: 278–89.

Marx, K. (1977) Critique of Hegel's Philosophy of Right. *Karl Marx: Selected Writings*. D. McLellan (ed.). New York, Oxford University Press: 27–30.

Marx, K. (1994) *Early Political Writings*. New York, Cambridge University Press.

Massey, D. (2005) *For Space*. Thousand Oaks, CA, Sage.

Mayer, M. (1994) Post-Fordist City Politics. *Post-Fordism: A Reader*. A. Amin (ed.). Malden, MA, Blackwell: 316–37.

Mayer, M. (2007) Contesting the Neoliberalization of Urban Governance. *Contesting Neoliberalism*. H. Leitner, J. Peck and E. Sheppard (eds). New York, Guilford: 90–115.

Meyer, D. and S. Staggenborg (1996) Movements, Countermovements, and the Structure of Political Opportunity. *American Journal of Sociology* 101(6): 1628–60.

Mill, J. (1976) M. De Tocqueville on Democracy in America. *John Stuart Mill on Politics and Society*. G. Williams (ed.). Brighton, Harvester Press: 186–247.

Mill, J. (1998[1859]) *On Liberty*. Oxford, Oxford University Press.

Miller, B. (2000) *Geography and Social Movements*. Minneapolis, University of Minnesota Press.

Miller, B. (2007) Modes of Governance, Modes of Resistance: Contesting Neoliberalism in Calgary. *Contesting Neoliberalism*. H. Leitner, J. Peck and E. Sheppard (eds). New York, Guilford: 223–49.

Mitchell, D. (2003) *The Right to the City: Social Justice and the Fight for Public Space*. New York, Guilford Press.

Modie, N. (2006) Port Opts to Exceed EPA's South Park Cleanup Plan: Recreational Use Could Be in Duwamish Tract's Future. *Seattle Post-Intelligencer*. June 28.

Mont Pelerin Society (1947) Statement of Aims. Mont Pelerin, Switzerland, Mont Pelerin Society.

Montgomery, D. (2005) An Indelible Moment: For Iraqi Nationals, Hope for Democracy Is Right at Their Fingertips. *Washington Post*. January 29, p. C1.

Moody, K. (1997) *Workers in a Lean World*. New York, Verso.

Mouffe, C. (1993) *The Return of the Political*. London, Verso.

Mouffe, C. (1995) Post-Marxism: Democracy and Identity. *Environment and Planning D: Society and Space* 13: 259–65.

Mouffe, C. (1996) Deconstruction, Pragmatism, and the Politics of Democracy. *Deconstruction and Pragmatism*. C. Mouffe (ed.). New York, Routledge: 1–13.

Mouffe, C. (1999) Deliberative Democracy or Agonistic Pluralism? *Social Research* 66(3): 745–58.

Mouffe, C. (2000) *The Democratic Paradox*. London, Verso.

Mouffe, C. (2002) Which Public Sphere for a Democratic Society. *Theoria* June: 55–65.

Mouffe, C. (2005) *On the Political*. New York, Routledge.

Mulady, K. (2004a) City's Ties to Vulcan Run Deep. *Seattle Post-Intelligencer*. October 25, p. A14.

Mulady, K. (2004b) Remaking South Lake Union: Seattle Is on Fast Track to Build Biotech Hub. *Seattle Post-Intelligencer*. October 20, p. A1.

Murphy, R. (2003) Nickels' Neighborhood Revolt. *Real Change*. June 12.

National Technology Transfer Center (2006) About NTTC. http://www.nttc.edu/, National Technology Transfer Center.

Norgaard, R. (1994) *Development Betrayed: The End of Progress and a Coevolutionary Revisioning of the Future*. New York, Routledge.

Nozick, R. (1974) *Anarchy, State, and Utopia*. New York, Basic Books.

Nylen, W. (2003) *Participatory Democracy Versus Elitist Democracy : Lessons from Brazil*. New York, Palgrave Macmillan.

Office of Emergency and Remedial Response (2005) Superfund Community Involvement Handbook. Washington, DC, USEPA.

Office of the Mayor (2003) Revitalized South Lake Union Would Create at Least 32,000 New Jobs, Report Says. Seattle, City of Seattle.

Office of the Mayor (2004a) Mayor Celebrates More Jobs in South Lake Union. Seattle, Office of the Mayor, June 16.

Office of the Mayor (2004b) State to Benefit from Development at South Lake Union. Seattle, Office of the Mayor, February 25.

Office of the President (2002) *The National Security Strategy of the United States of America*. Washington, DC, The White House.

Office of the President (2006) *The National Security Strategy of the United States of America*. Washington, DC, The White House.

Olds, K. (1998) Urban Mega-Events, Evictions and Housing Rights: The Canadian Case. *Current Issues in Tourism* 1(1): 2–46.

Oliver, M. and T. Shapiro (1997) *Black Wealth/White Wealth*. New York, Routledge.

O'Neill, K. (1992) *Detached Communities: The Segregationist Effect of Mandatory Homeowner Associations*. Masters Thesis, Department of Urban Planning, UCLA.

Parker, W. (2003) *Teaching Democracy: Unity and Diversity in Public Life*. New York, Teachers College Press.

Pateman, C. (1970) *Participation and Democratic Theory.* Cambridge, Cambridge University Press.

Pateman, C. (1985) *The Problem of Political Obligation: A Critique of Liberal Theory.* Berkeley, University of California Press.

Pateman, C. (1987) Feminist Critiques of the Public/Private Dichotomy. *Feminism and Equality.* A. Phillips (ed.). New York, New York University Press: 103–26.

Payne, T. and C. Skelcher (1997) Explaining Less Accountability: The Growth of Local QUANGOs. *Public Administration* 75(2): 207–24.

Peck, J. (1998) Geographies of Governance: TECs and the Neo-Liberalisation of "Local Interests". *Space & Polity* 2(1): 5–31.

Peck, J. (2001) *Workfare States.* New York, Guilford.

Peck, J. and A. Tickell (1994) Searching for a New Institutional Fix: The *after*-Fordist Crisis and the Global–Local Disorder. *Post-Fordism: A Reader.* A. Amin (ed.). Malden, Blackwell: 280–315.

Peck, J. and A. Tickell (2002a) Neoliberalising Space. *Antipode* 34(3): 380–404.

Peck, J. and A. Tickell (2002b) Neoliberalizing Space. *Spaces of Neoliberalism.* N. Brenner and N. Theodore (eds). Malden, Blackwell: 33–57.

Peterson, P. (1981) *City Limits.* Chicago, University of Chicago Press.

Petit, P. (1997) *Republicanism: A Theory of Freedom and Government.* Oxford, Oxford University Press.

Pincetl, S. (1994) Challenges to Citizenship: Latino Immigrants and Political Organizing in the Los Angeles Area. *Environment and Planning A* 26(6): 895–914.

Piore, M. and C. Sabel (1984) *The Second Industrial Divide.* New York, Basic Books.

Piven, F. and R. Cloward (2000) Power Repertoires and Globalization. *Politics and Society* 28: 413–30.

Pollin, R. (2003) *Contours of Descent.* New York, Verso.

Polsby, N. (1980) *Community Power and Political Theory.* New Haven, Yale University Press.

Polybius (1979) *The Rise of the Roman Empire.* New York, Penguin.

Ponnuru, R. (1999) "We Are All Clueless Now": The Eclipse of Economics—Neither Party Has a Clear Economic Policy. *National Review,* November 8.

Port of Seattle (2000) Partnership Forms to Study Lower Duwamish River. Seattle, Port of Seattle.

Post-Intelligencer Staff (2007) The Political History of the Viaduct. *Seattle Post-Intelligencer.* March 14.

Puget Sound Regional Council (2005) Puget Sound Trends. Seattle, Puget Sound Regional Council.

Purcell, M. (2001a) Metropolitan Political Reorganization and the Political Economy of Urban Growth: The Case of San Fernando Valley Secession. *Political Geography* 20(5): 101–21.

Purcell, M. (2001b) Neighborhood Activism among Homeowners as a Politics of Space. *Professional Geographer* 53(2): 178–94.

Purcell, M. (2002a) Excavating Lefebvre: The Right to the City and Its Urban Politics of the Inhabitant. *GeoJournal* 58(2–3): 99–108.

Purcell, M. (2002b) The State, Regulation, and Global Restructuring: Reasserting the Political in Political Economy. *Review of International Political Economy* 9(2): 284–318.

Purcell, M. (2003a) Citizenship and the Right to the Global City: Reimagining the Capitalist World Order. *International Journal of Urban and Regional Research* 27(3): 564–90.

Purcell, M. (2003b) Islands of Practice and the Marston/Brenner Debate: Toward a More Synthetic Critical Human Geography. *Progress in Human Geography* 27(4): 317–32.

Purcell, M. (2006) Urban Democracy and the Local Trap. *Urban Studies* 43(11): 1921–41.

Purcell, M. and J. C. Brown (2005) Against the Local Trap: Scale and the Study of Environment and Development. *Progress in Development Studies* 5(4): 279–97.

Purcell, M. and J. Nevins (2005) Pushing the Boundary: State Restructuring, Regulation Theory, and the Case of U.S.–Mexico Border Enforcement in the 1990s. *Political Geography* 24(2): 211–35.

Rawls, J. (1993) *Political Liberalism*. New York, Columbia University Press.

Rights to the City (1998) Conference Held at York University, Toronto, Canada. June.

Rights to the City (2002) Conference Held in Rome, Italy. May.

Riley, J. (1988) *Liberal Utilitarianism: Social Choice Theory and J.S. Mill's Philosophy*. Cambridge, Cambridge University Press.

Rivera, D. (2004) Portland, Ore., Developer Finds Seattle More Receptive Climate for Projects. *The Oregonian*. October 24.

Robertson, R. (1995) Glocalization: Time-Space and Homogeneity-Heterogeneity. *Global Modernities*. M. Featherstone, S. Lash and R. Robertson (eds). London, Sage: 91–107.

Rousseau, J. (1987) *Basic Political Writings*. Indianapolis, Hackett.

Royale, R. (2006) Toxic Avengers: Duwamish Watchdogs Want Port to Be Diligent in Clean up of PCBs. *Real Change*. November 16.

Rubin, H. and I. Rubin (2007) *Community Organizing and Development*. Boston, Allyn & Bacon.

Sabel, C. (1994) Flexible Specialization and the Re-Emergence of Regional Economies. *Post-Fordism: A Reader*. A. Amin (ed.). Malden, Blackwell: 101–56.

Sack, R. (1993) The Power of Space and Place. *The Geographical Review* 83: 326–9.

Salmon, S. (2001) The Right to the City? Globalism, Citizenship, and the Struggle over Urban Space. 97th Annual Meeting of the Association of American Geographers, New York, February.

Samara, T. (2007) Right to the City: Notes from the Inaugural Convening. Los Angeles, Strategic Actions for a Just Economy.

Sandel, M. (1996) *Democracy's Discontent: America in Search of a Public*. Cambridge, MA, Harvard University Press.

Sandercock, L. (1998) The Death of Modernist Planning: Radical Praxis for a Postmodern Age. *Cities for Citizens: Planning and the Rise of Civil Society in a Global Age*. J. Friedmann and M. Douglass (eds). New York, John Wiley & Sons: 163–84.

Sanders, L. (1997) Against Deliberation. *Political Theory* 25(3): 347–76.

Sanga, R. (2006) T-117 Early Action Memorandum. Seattle, USEPA Region 10.

Sartori, G. (1987) *The Theory of Democracy Revisited*. Catham, Catham Publishers.

Sassen, S. (1994) *Cities in a Global Economy*. Thousand Oaks, Pine Forge Press.

Sassen, S. (1999) Whose City Is It? Globalization and the Formation of New Claims. *Cities and Citizenship*. J. Holston (ed.). Durham, Duke University Press: 177–94.

Sassen, S. (2000) The Global City: Strategic Site/New Frontier. *Democracy, Citizenship and the Global City*. E. Isin (ed.). New York, Routledge: 48–61.

Sato, M. (1997) *The Price of Taming a River.* Seattle, Mountaineers Books.

Saxenian, A. (1994) *Regional Advantage: Culture and Competition in Silicon Valley and Route 128.* Cambridge, Harvard University Press.

Schmitt, C. (1971) *The Concept of the Political.* New Brunswick, Rutgers University Press.

Schoenbrod, D. (1996) Why States, Not EPA, Should Set Pollution Standards. *Regulation: The Review of Business and Government* 19(4).

Schumpeter, J. (1947) *Capitalism, Socialism, and Democracy.* New York, Harper and Brothers.

Scott, A. (1988) *New Industrial Spaces.* London, Pion.

Scott, A. (1993) *Technopolis: High-Technology Industry and Regional Development in Southern California.* Los Angeles, University of California Press.

Scott, A. (1996) Regional Motors of the Global Economy. *Futures* 28(5): 391–411.

Scott, A. (1998) *Regions and the World Economy: The Coming Shape of Global Production, Competition and Political Order.* New York, Oxford University Press.

Scott, A. (2006) Port OKs $6 Million to Start Cleanup of South Park Site. *Seattle Times.* June 28, p. E1.

Sites, W. (2007) Contesting the Neoliberal City? Theories of Neoliberalism and Urban Strategies of Contention. *Contesting Neoliberalism.* H. Leitner, J. Peck and E. Sheppard (eds). New York, Guilford: 116–38.

Skinner, Q. (1992) On Justice, the Common Good, and the Priority of Liberty. *Dimensions of Radical Democracy: Pluralism, Citizenship, Community.* C. Mouffe (ed.). London, Verso: 211–24.

Smith, N. (1984) *Uneven Development: Nature, Capital and the Production of Space.* New York, Basil Blackwell.

Smith, N. (1987) Dangers of the Empirical Turn: Some Comments on the CURS Initiative. *Antipode* 19: 59–68.

Smith, N. (1993) Homeless/Global: Scaling Places. *Mapping the Futures: Local Cultures Global Change.* J. Bird (ed.). New York, Routledge: 87–119.

Smith, N. (1995) Remaking Scale: Competition and Cooperation in Prenational and Postnational Europe. *Competitive European Peripheries.* H. Eskelinen and F. Snickars (eds). Berlin, Springer Verlag: 59–74.

Smith, N. (1996) *The New Urban Frontier: Gentrification and the Revanchist City.* New York, Routledge.

Smith, N. (2003) Forward to Henri Lefebvre's. *The Urban Revolution.* Minneapolis, University of Minnesota Press: vii–xxiii.

Soja, E. (1980) The Socio-Spatial Dialectic. *Annals of the Association of American Geographers* 70(2): 207–25.

Soja, E. (1989) *Postmodern Geographies.* London, Verso.

Soja, E. (1996) *Thirdspace: Journeys to Los Angeles and Other Real-and-Imagined Places.* Cambridge, MA, Blackwell.

Soja, E. (2000) *Postmetropolis.* Malden, Blackwell.

Staeheli, L., J. Kodras and C. Flint (eds) (1997) *State Devolution in America: Implications for a Diverse Society.* Thousand Oaks, Sage.

Stiglitz, J. (2004) Cancún Failure: A Triumph for Democracy. *South Bulletin* 79: 230–5.

Storper, M. (1997) *The Regional World: Territorial Development in a Global Economy.* New York, Guilford.

Students for a Democratic Society (1962) *The Port Huron Statement of the Students for a Democratic Society*. Port Huron, MI, Students for a Democratic Society.

Sunstein, C. (1997) Deliberation, Democracy, and Disagreement. *Justice and Democracy: Cross-Cultural Perspectives*. R. Bontekoe and M. Stepaniants (eds). Honolulu, University of Hawai'i Press: 93–117.

Susskind, L. and S. Podziba (1999) Affordable Housing Mediation: Building Consensus for Regional Agreements in the Hartford Area. *The Consensus-Building Handbook: A Comprehensive Guide to Reaching Agreement*. L. Susskind, S. McKearnan and J. Thomas-Larner (eds). Thousand Oaks, Sage: 773–800.

Susskind, L., S. McKearnan and J. Thomas-Larner (eds) (1999) *The Consensus Building Handbook*. Cambridge, MA, MIT Press.

Swyngedouw, E. (1992) The Mammon Quest. "Glocalization," Interspatial Competition and the Monetary Order: The Construction of New Spatial Scales. *Cities and Regions in the New Europe: The Global–Local Interplay and Spatial Development Strategies*. M. Dunford and G. Kafkalas (eds). London, Belhaven: 39–67.

Swyngedouw, E. (1996) Reconstructing Citizenship, the Re-Scaling of the State and the New Authoritarianism: Closing the Belgian Mines. *Urban Studies* 33(8): 1499–521.

Swyngedouw, E. (1997) Neither Global nor Local: "Glocalization" and the Politics of Scale. *Spaces of Globalization*. K. Cox (ed.). New York, Guilford Press: 137–66.

Swyngedouw, E., F. Moulaert and A. Rodriguez (2002) Neoliberal Urbanization in Europe: Large-Scale Urban Development Projects and the New Urban Policy. *Antipode* 34(3): 542–77.

Tabb, W. (1982) *The Long Default: New York City and the Urban Fiscal Crisis*. New York, Monthly Review Press.

Tam, H. (1998) *Communitarianism: A New Agenda for Politics and Citizenship*. New York, New York University Press.

Tarrow, S. (1996) States and Opportunities: The Political Structuring of Social Movements. *Comparative Perspectives on Social Movements*. D. McAdam, J. McCarthy and M. Zald (eds). New York, Cambridge University Press: 41–61.

Task Force on Inequality and American Democracy (2004) *American Democracy in an Age of Rising Inequality*. Washington DC, American Political Science Association.

Taylor, C. (1992) *Multiculturalism and "the Politics of Recognition"*. Princeton, Princeton University Press.

Tewdwr-Jones, M. and P. Allmendinger (1998) Deconstructing Communicative Rationality: A Critique of Habermasian Collaborative Planning. *Environment and Planning A* 30: 1975–89.

Thatcher, M. (1987) Interview with Margaret Thatcher. *Woman's Own*. October 31.

Thrift, N. (1996) *Spatial Formations*. Thousand Oaks, Sage.

Tickell, A. and J. Peck (2003) Making Global Rules: Globalisation or Neoliberalisation? *Remaking the Global Economy: Economic-Geographical Perspectives*. J. Peck and H. Yeung (eds). London, Sage: 163–82.

Tilly, C. (1984) Social Movements and National Politics. *Statemaking and Social Movements: Essays in History and Theory*. C. Bright and S. Harding (eds). Ann Arbor, University of Michigan Press: 297–317.

Tindall, M. (2004) Urban Redevelopment Spurs Economic Growth. *Business Xpansion Journal*. May 1.

Torvik, S. (2000) Duwamish Proposed as Superfund Site. *Seattle Post-Intelligencer*. December 6.

Traub, J. (2006) Bad Company. *The New York Times Book Review*. August 13, p. 7.

Tullock, G. (1970) *Private Wants, Public Means: An Economic Analysis of the Desirable Scope of Government*. New York, Basic Books.

Tushnet, M. (1984) An Essay on Rights. *Texas Law Review* 62: 1363–412.

UNESCO (2006) *International Public Debates: Urban Policies and the Right to the City*. Paris, UNESCO.

United Nations Center for Human Settlements (2001) Policy Dialogue Series: Number 1 Women and Urban Governance. New York, UNCHS (Habitat).

USEPA and Washington State Department of Ecology (2003) Fact Sheet: Lower Duwamish Waterway Site. June, Seattle, USEPA Region 10.

van den Brink, B. (2005) Liberalism without Agreement: Political Autonomy and Agonistic Citizenship. *Autonomy and the Challenges to Liberalism*. J. Christman and J. Anderson (eds). New York, Cambridge University Press: 245–71.

Van Dyk, T. (2005) Let's Detour from the Seattle Way. *Seattle Post-Intelligencer*. January 6.

Van Zandt, T. (1971) Tower Song. *Delta Momma Blues*. New York, Tomato Records.

Vaughan-Nichols, S. (2006) Let the Browser Wars Begin. *DesktopLinux.com*. October.

Verba, S. and N. Nie (1972) *Participation in America: Political Democracy and Social Equality*. New York, Harper & Row.

Vulcan Real Estate (2007) The S.L.U. Story. Seattle, Vulcan, Inc., retrieved from http://www.vulcanrealestate.com/TemplateSouthLakeUnion.aspx?contentId=11.

Wainwright, J. (2007) Spaces of Resistance in Seattle and Cancun. *Contesting Neoliberalism*. H. Leitner, J. Peck and E. Sheppard (eds). New York, Guilford: 179–203.

Wallach, L. and M. Sforza (1999) *Whose Trade Organization? Corporate Globalization and the Erosion of Democracy*. Washington, DC, Public Citizen.

Walzer, N. and L. York (1998) Public–Private Partnerships in U.S. Cities. *Public-Private Partnerships for Local Economic Development*. N. Walzer and B. Jacobs (eds). Westport, Praeger: 47–67.

Ward, K. (2000) A Critique in Search of a Corpus: Re-Visiting Governance and Re-Interpreting Urban Politics. *Transactions of the Institute of British Geographers* 25(2): 169–85.

Washburn, J. (2005) *University, Inc.: The Corporate Corruption of American Higher Education*. New York, Basic Books.

Washington Competitiveness Council (2002) Keeping Washington's Competitive Edge. Seattle, Washington Competitiveness Council.

Washington Competitiveness Council (2003) 2003 Legislative Session Highlights. Seattle, Washington Competitiveness Council.

Washington Competitiveness Council (2004) 2004 Legislative Session Highlights. Seattle, Washington Competitiveness Council.

Washington State Department of Health (2003) Public Health Assessment: Lower Duwamish Waterway. Olympia, WA, Washington State Department of Health.

Watson, S. (1995) Using Public–Private Partnerships to Develop Local Economies: An Analysis of Two Missouri Enterprise Zones. *Policy Studies Journal* 23(4): 652–67.

Whitman, W. (2004) *Complete Prose Works*. Whitefish, Kessinger.

Wilson, D. (2004) Making Historical Preservation in Chicago: Discourse and Spatiality in Neo-Liberal Times. *Space and Polity* 8(1): 43–59.

Wittgenstein, L. (1953) *Philosophical Investigations*. Oxford, Basil Blackwell.

Wolin, S. (1994) Fugitive Democracy. *Constellations* 1(1): 11–25.

Worldwide Conference on the Right to Cities Free From Discrimination and Inequality (2002) Porto Allege, Brazil. February.

Young, B. (2003) Zoning Plan for South Lake Union Sparks Debate. *Seattle Times.* September 24.

Young, I. (1990) *Justice and the Politics of Difference.* Princeton, Princeton University Press.

Young, I. (1996) Communication and the Other: Beyond Deliberative Democracy. *Democracy and Difference.* S. Benhabib (ed.). Princeton, Princeton University Press: 120–36.

Young, I. (1999a) Difference as a Resource for Democratic Communication. *Deliberative Democracy.* J. Bohman and W. Rehg (eds). Boston, MIT Press: 383–406.

Young, I. (1999b) Residential Segregation and Differentiated Citizenship. *Citizenship Studies* 3(2): 237–52.

Young, I. (2000) *Inclusion and Democracy.* New York, Oxford University Press.

Young, I. (2001) Activist Challenges to Deliberative Democracy. *Political Theory* 29(5): 670–90.

Yuval-Davis, N. (1997) Women, Citizenship and Difference. *Feminist Review* 57: 4–28.

Yuval-Davis, N. (1999) "The Multi-Layered Citizen": Citizenship at the Age of "Glocalization". *International Feminist Journal of Politics* 1(1): 119–36.

Zizek, S. (1992) *Enjoy Your Symptom.* London, Routledge.

INDEX

Agnew, J. xii, 9
agonism/antagonism 66–7, 73, 80–3,
    104–5, 134, 154, 183
agreement (vs. conflict) *see* conflict
Appiah, K. 40, 41, 43
Aristotle 52–4, 105, 178, 184

Barber, B. 53
Benhabib, S. 45, 46, 57
Berlin, I. 41, 44
Bowles, S. and H. Gintis 10, 35, 87, 88
Brazil 1, 82; participatory budgeting in
    26, 55–6, 97; and right to the city 96–8
Brenner, N. xi, xii, 9, 13, 15, 16, 21, 26, 89,
    102, 183
Brown, J.C. xii, 101, 162
Brown, M. xii, 185
Bush, G.W. 31, 33–4, 37, 169

Castells, M. 89, 94
chains of equivalence *see* equivalence
city/cities 88–90, 102–3, 177–9
Cohen, J. 44, 46, 47–8, 55,
common good (vs. particular interests)
    26, 39, 43, 45–6, 50, 51, 52, 53–4, 56–7,
    67–70, 73–82, 96, 104, 111, 184
communicative action/rationality 44–51,
    56, 69–72, 77, 79, 126
communicative/collaborative planning
    29, 46, 48–9, 80, 126

conflict (vs. agreement) 8, 45, 47, 48–9,
    56, 62–3, 66, 67–70, 73, 77, 111, 144,
    154–5
Connolly, W. 62–3, 68, 71
consensus 47, 48–9, 56, 67, 69, 73, 78,
    79–80, 81, 111, 134, 154
cultural esteem/recognition 50, 72–3,
    85, 86
Cunningham, F. 42, 53

Dahl, R. 62–3, 84
Davis, M. xii, 23, 96, 122, 156–7
Day, R. 38, 85
Deleuze, G. and F. Guattari 185
democracy/democratization 33–74;
    attitudes 3–5, 37–8, 61, 76–88, 90, 96,
    98–9, 104–5, 108, 123–4, 154, 176;
    citizens/citizenship 2, 19, 21, 24, 25,
    26–7, 30, 39, 42, 44, 46–7, 52–5, 64,
    86, 93, 96, 129–30, 170, 173, 184; civic
    republicanism 41, 52, 57, 83, 184;
    classic pluralism 62–3, 67; **deliberative**
    26, 28, 39, 44–51, 53–4, 56, 61, 62,
    64, 67–73, 78, 80, 111, 125–35, 154,
    156–7, 182; direct 25, 35, 184; equality
    25–6, 39, 44–5, 50, 58, 62, 64–5, 72–3,
    77–81, 83–7, 95, 97, 105, 155, 170,
    173; elections/representation 30,
    33–5, 42, 53–4, 58–9, 65, 111, 170;
    inclusion/exclusion 45, 50, 68–9, 71–2,

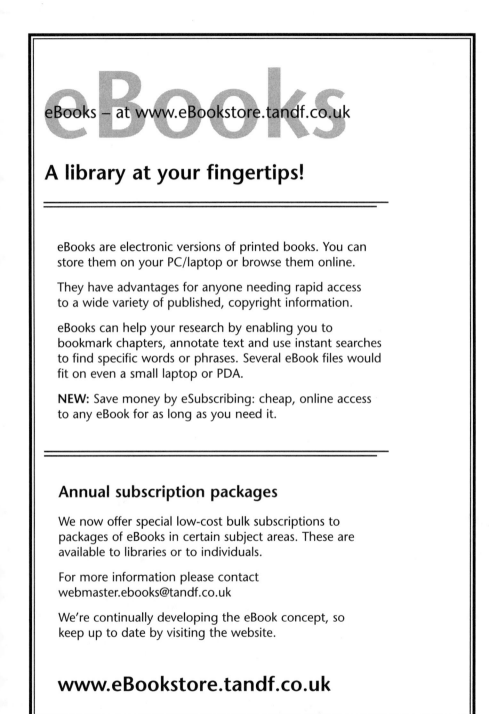